How to Catch a Robot Rat

How to Catch a Robot Rat

When Biology Inspires Innovation

Agnès Guillot and Jean-Arcady Meyer

Translated by Susan Emanuel

The MIT Press
Cambridge, Massachusetts
London, England

Originally published in French as *La bionique: Quand la science imite la Nature* by Dunod, 2008, Paris.

For information about special quantity discounts, please email special_sales @mitpress.mit.edu

This book was set in Stone Sans and Stone Serif by Toppan Best-set Premedia Limited.

Printed and bound in the United States of America.

Library of Congress Cataloging-in-Publication Data
Guillot, Agnès, 1946–
[Bionique. English]
How to catch a robot rat : when biology inspires innovation / Agnès Guillot and Jean-Arcady Meyer ; translated by Susan Emanuel.
    p. cm.
Includes bibliographical references and index.
ISBN 978-0-262-01452-6 (hardcover : alk. paper) 1. Bionics. I. Meyer, Jean-Arcady. II. Title.
Q320.G7813   2010
600—dc22

                                                              2009054313

10 9 8 7 6 5 4 3 2 1

To Agathe,
So that she may take care of this planet
On which so many marvels are evolving

Contents

Conclusion   199

Epilogue   207

Acknowledgments

We would like to thank all our colleagues who agreed to reread certain paragraphs or chapters to catch the inevitable mistakes, inaccuracies, or difficulties in phrasing.

We also thank all those who devoted their precious time to provide the references and illustrations that we wanted to include in this book.

Special thanks go to Jean Solé, whose drawings have often magnificently illustrated our efforts to promote the biomimetic approach to robotics.

Our special gratitude goes to Angelo Arleo, Kellar Autumn, Wilhelm Barthlott, Serge Berthier, Aude Billard, Julie Blackburn, Rodney Brooks, Roberto Cordeschi, Thomas DeMarse, Dennis Dollens, Stéphane Doncieux, Kenji Doya, Cyrille Foasso, Stéphanie Georget, Benoît Girard, Stanislav Gorb, Rex Graham, Owen Holland, Ioannis Ieropoulos, Auke Jan Ijspeert, Caroline Junier, Frédéric Kaplan, Mehdi Khamassi, Jacqueline Koeppen, Francis Lara, Jacques Maigret, Chris Melhuish, Alain Mercier, Christophe Meyer, Robert Michelson, Xavier de Montfort, Francesco Mondada, René Motro, Jean-Baptiste Mouret, Naohisa Nagaya, Rolf Pfeifer, Patrick Pirim, Giulio Sandini, Kim Sangbae, Isabelle Taillebourg, Henri Van Damme, and Stuart Wilkinson . . .

. . . as well as to all the other animals and plants of this planet that have found adaptations so ingenious as to subsist here.

Agnès Guillot and Jean-Arcady Meyer

Introduction

Nature is a universal and certain teacher of those who observe it.
—Carlo Goldoni (1750)

A fiber inspired by a spider web, five times more resistant than steel and twice as elastic as nylon; a microdrone beatings its wings like an insect; a cyberhand that gives its wearer back the sensations of touch and warmth . . . whatever the evolution of species has dared to imagine, humankind has tried to reproduce it!

Testimony of such an enterprise—historical or legendary—has been recorded since the invention of writing. It is said that, twenty-four centuries ago, a Greek man apparently conceived of a wooden dove capable of flying; three centuries later the Egyptians supposedly fabricated an articulated artificial hand; at the start of our era, a Chinese man invented paper by observing a wasps' nest. In every age, human intelligence has delved into the intelligence of nature in order to perfect its own inventions, even if sometimes the technical means to realize them does not exist: the flying machines inspired by birds and bats that Leonardo da Vinci sketched could not take off for want of materials sufficiently light and motors sufficiently powerful.

Since the nineteenth century, though, technological progress has been astounding. Encouraged by the Industrial Revolution, engineers have invented new means to realize products and machines. In our day, in particular, another revolution is under way: the exploration of the infinitely small, with the manipulation of matter at the level of the atom on the order of a billionth of a meter. Nanotechnologies foretell a field of entirely original applications. Apart from the advantage of miniaturization through nanotechnologies, the mechanical, electrical, and optical characteristics of the materials produced with them are radically different from characteristics we can observe in our macroworld. Artifacts having the same microscopic structures as those of the living world thus become conceivable by researchers.

In parallel to these technical developments, new currents of thought have brought natural and artificial systems closer together. In the 1940s and 1950s, psychologist Clark Leonard Hull argued that identical laws govern the behavior of animals and machines and that the latter are perfectly capable of learning. At the same time, mathematician Norbert Wiener founded a new discipline—which he called "cybernetics"—centered on the analogy between data processing performed by natural systems and data processing performed by artificial systems. The pioneers of artificial intelligence were soon exploiting this kind of comparison and went so far as to propose an equivalence between the computer and the brain. From this perspective, it is the artificial that becomes a metaphor for the living and not the reverse.

The idea that the life sciences and engineering sciences might mutually reinforce each other took shape, and in September 1960 an American conference organized at Dayton, Ohio, popularized the term *bionics*,[1] a contraction of *biology* and *techniques*, and later of *biology* and *electronics*, and laid out the scientific objectives of this discipline. The conference aimed to launch a vast program uniting the knowledge and skills of engineers and researchers from various backgrounds—mathematicians, physicists, and chemists, but also biologists and psychologists—with a view to seeking in nature how to conceive and produce effective artificial systems. Ten years later, what we could call the "new bionics" would extend this research to the conception of hybrid systems that integrate living components with artificial components—as popularized in the well-known American television series *The Six Million Dollar Man*.

This book devoted to the new bionics has three sections. The first describes some of the many technological achievements inspired by natural structures, processes, and materials. For example, we look at how the study of the way in which the gecko's feet adhere for 1/8,000 of a second to any surface has led to the resolution of the acute problem of designing a dry adhesion process that is perfectly repositionable independently of the support involved. The second section explains one field of research (that is of ancient inspiration but has been prospering for a few years): the concept of autonomous robots inspired by animals and their behavior—what is commonly called "bioinspired robotics." For example, we show how the way in which an albatross coasts for whole days with hardly any wing beats may bring about a solution to the problem of how future microdrones might carry small fuel tanks. Finally, the third section examines work aiming to hybridize natural and artificial systems; we look at how "neuroprostheses" may allow quadriplegic patients to control machines at a distance.

We develop many other examples of invention and hybridization, although we cannot be exhaustive because research in the subject has accelerated so much. Our historical account of the individual evolution of these various domains of application is also brief. At the end of the book, we mention some perspectives—at least those imagined by a few adventurous researchers. We also allude to some limits this kind of work is confronted with and the ethical problems that might arise, as indicated by the development of various legal texts and charters that aim to prevent the risk of any harmful use of these artificial inventions.

I Structures, Processes, Materials

1 Nature-Technology Transfers: From Artisanry to Industry

The Genius of Man can produce numerous inventions thanks to the application of various instruments contributing to the same goal. Yet, he will make none more beautiful, simple, or better adapted than those of Nature, for, in her own inventions, nothing is lacking and nothing is superfluous.
—Leonardo da Vinci (1480)

Well before being a science, bionics was practiced at an artisanal level to improve daily life or to try to augment human capabilities. In the nineteenth century, individual patents protecting these inventions began to appear, and industries did not take long to begin exploiting these inventions on a greater scale.

Inventions for Daily Life

Eight centuries ago in China, the villagers of Hongcun in the Huangshan mountains may have been the first bionic architects. They gave their village the form of an ox and conceived of a sophisticated hydrological network shaped like a digestive system for the transport of water. A hill represented the head, and two large trees were the horns. Four bridges formed the feet, and the houses of the village were the body. The channels that brought fresh water to each house were the intestines, which ended in a lake-stomach shaped like a half moon, then pursued their course to a larger lake—the belly of the beast—before tumbling into a river. In our day, this system can still bring drinkable water and irrigate village crops, even if the villagers now use it more for touristic than practical purposes, especially because Hongcun has been classified as a UNESCO world site.

This example arises more from symbolic interpretation—the ox being the sign of longevity in China—than from the patent transfer of inventions born of the evolution of species. The origins and inspiration of other achievements are more difficult to discern. It is claimed that artificial rock dams dating from 3,000 BCE were inspired by raccoons' dams and that primitive fishermen might have examined the network of spider webs to weave similar webs for their nets. In the same way, the efficient thermoregulation of houses that are like cavities made of clay mixed with water—or adobe—built by the Pueblo Indians of New Mexico may have resulted from their imitating the architecture of solitary wasps' clay nests.

No discoverer has left a name on these achievements, but others had more luck with publicity. The Chinese Ts'ai-Lun is recognized as the inventor of paper, a material that he produced in 105 CE from a mixture of bamboo and water, supposedly after having carefully observed the nest of social wasps, the cousins of the solitary wasps. The nest of these hymenoptera is made of wood mixed with saliva and organized in strata of cells linked by pillars, the whole enclosed in several layers of the same material. It is also said that much later (1719) the same kind of observation led the French physicist René-Antoine Ferchault de Réaumur to propose the current recipe for fabricating paper as a replacement for the less rigid rag paper used until then.

Industrial Advances

Groping around in their more or less fruitful attempts, most inventors by the nineteenth century were filing patents for their apparatuses. The artisanal approach had reached an end.

The history of barbed wire is a good example of the shift from the artisanal era to the industrial era—and of the harmful consequences this shift could entail. In the American West, this iron wire at first served as a substitute for the branches of a thorny tree from Texas, the Osages orange. The Wazházhe—a Sioux name mispronounced by French missionaries as "Osage"—made hoops with this tree's branches and used the latex of the tree's inedible fruit to paint their faces. To keep animals, the first American colonists of the Southwest had the idea of surrounding their land with a cordon of these orange trees, for the thorny hedges, they said, were "high enough for horses, resistant to cows, and thick enough for pigs." Then, to compensate for these natural barriers' very slow growth, a few farmers were astute enough to replace them with metallic wires wrapped at intervals with iron barbs imitating the prickly branches. Nicknamed "the devil's cords," these wires were perfected by Joseph Farewell Glidden, who in 1874 filed a patent for barbed wire, which a merchant of his town soon put into production.

This invention, at the start very convenient, had dramatic consequences. Cowboys lost their jobs, becoming unemployed due to technology. Barbed-wire boundaries, having become removable and extensible, aroused bitter fights among farmers to win patches of land. They also allowed farmers and ranchers to extend their possessions, thereby pushing the Indian tribes and their principal nourishment—the bison—into increasingly narrow spaces. Then the Civil War turned barbed wire into a weapon by using it to trap assailants and to store prisoners, a weapon that has been characterized as "a tool for the political management of space."[1]

Times of war are unfortunately propitious for the recuperation and accelerated use of any innovation. Combatants in the First World War exploited sonar for underwater detection, perfected by the French physicist Paul Langevin. But let us put this example to one side. Although it might seem directly inspired by aquatic animals' use of echolocation, that was not the case: the convergence between the two processes was noticed only much later.

In contrast, at the time of the Second World War, the same Paul Langevin was curiously involved in an industrial application that was incontestably inspired by biology, the Cricket Watch, the first bracelet watch with an alarm function. The American army was looking to furnish its soldiers with a gadget that would allow them to synchronize their operations. The Swiss firm Vulcain had already produced a watch alarm around 1890, but with two serious defects: the vibrations caused by the alarm systematically deregulated the time, and the sound of this alarm was easily smothered by ambient noise. Engineers strove to improve this apparatus but began to be discouraged, until one day Langevin, on a visit to Vulcain, asserted that the problem was indeed solvable: if a cricket, a minuscule insect, could produce a sound carrying several meters, then a watch could obviously do so. . . . Thus was born the idea of placing a steel membrane that a tiny hammer made resonate inside a dual base pierced with holes to amplify the sound produced, on the model of the insect's stridulations. The Cricket Watch was still perfectly up to date in 1947, when it became, after Eisenhower wore one, the fetish watch of several American presidents.

In 1960, the Cold War was building to a fever pitch: the Soviets shot down a U-2 American plane and imprisoned the pilot. This era saw the famous founding of bionics at the Dayton conference organized by the U.S. Air Force at the Wright-Patterson base. Beyond the secret military applications to which this conference gave rise, it had a much wider impact and inspired research by a large number of public and private laboratories throughout the world. Thus, a first scientific symposium on bionics was organized in 1966 in Italy, focusing on the sonar of terrestrial and marine animals. Since then, research into bionics has grown enormously around the world. In 2005, a whole German pavilion was devoted to bionics at the universal exhibition in Aichi, Japan.

Chapters 2, 3, and 4 illustrate the dynamism of bionic technological inventions inspired by living systems. Living systems have experimented (and continue to experiment) with numerous morphologies and numerous mechanisms liable to ensure their survival. After two and a half billion years during which isolated cells had proved their extraordinary resistance in an aquatic milieu, a multitude of multicellular organisms with surrealist forms emerged.[2] Out of these countless architectural attempts, only a few representatives have remained, but they are sufficiently equipped to subsist still in today's marine environment.

When cyanobacteria and plants first became bold enough to exploit a new ecological niche, terra firma, they had to modify their structures and

Figure 1.1
A few examples of Cambrian fauna with "improbable" morphologies. Most of them belong to groups that are now totally extinct; out of 120 genera, approximately 33 remain, including sponges and mollusks, for example. (© Dunod.)

their internal mechanisms to fight against dryness, weight, ultraviolet rays, and abrupt changes in climatic conditions. Invertebrate animals and then the vertebrates soon followed the bacteria and plants onto land. Their morphology and their physiology were transformed under the same pressures that changed the vegetals.

Thus, life invented a large panoply of equipment adapted to an enormous variety of modes of survival on the planet. With what basic ingredients, though? Always the same! About 99 percent carbon, hydrogen, oxygen, and nitrogen, with the remainder being composed of twenty-some other elements out of the hundred or so that compose our known universe.

But nature does not easily give up its secrets regarding how things are made.

2 Structures

To copy the ever-open great book of nature.
—Antonio Gaudí (1883)

The Roman architect Vitruvius set the tone 2,100 years ago when he decreed that a building "must have an exact proportion worked out after the fashion of the members of a finely shaped human body." In other words, architecture should be inspired by the organic world, especially by what is most perfect, the human body.

This view lasted until the eighteenth century, when natural models were enlarged to include other vegetable and animal creatures, according to the idea that artificial structures are subject to the same physical laws as natural structures.

Armatures

The leaf of the water lily *Victoria amazonica* can reach two meters in diameter. Its rigid radiating veins growing from the stem are considerably strengthened by multiple, flexible concentric veins arranged in an orthogonal direction. These armatures render the leaf so solid that one leaf was able to support the weight of Annie, the little daughter of British architect Joseph Paxton.

A gardener at first, Paxton was named superintendent of the grounds of Chatsworth House, where he tried architecture. In a greenhouse there, he grew a cutting from the water lily *Victoria*, found in Guyana, which multiplied and extended so vertiginously that he was obliged to envision another construction to protect his cultivations. He thus conceived of an edifice of glass and iron, with armatures similar to those of this plant's leaf: the "Lily House" was very successful.

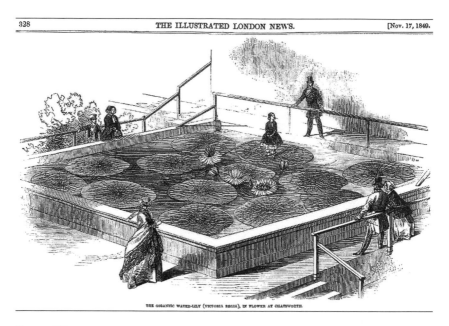

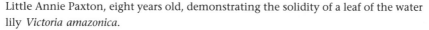

THE GIGANTIC WATER-LILY (VICTORIA REGIA), IN FLOWER AT CHATSWORTH.

Figure 2.1

Little Annie Paxton, eight years old, demonstrating the solidity of a leaf of the water lily *Victoria amazonica*.

This structure also inspired Paxton to design (in only nine days) the Crystal Palace in London, the star attraction of the Universal Exhibition of 1851. This revolutionary construction was entirely modular, prefabricated in 293,000 glass panels that two thousand men assembled in only eight months. Light but very resistant, it stood for 140 years as the site of many spectacular exhibitions before being destroyed in a fire. Constructed on the same model as the Exhibition of the Industry of All Nations of 1853, the Crystal Palace of New York curiously suffered the same fate: it was entirely destroyed three years later in a great fire.

Animal skeletons are also good models. Whether the arrangement of the four pillars of the Eiffel Tower were inspired by the internal structure of our femur, which is articulated at a slant from our hip, remains controversial. Nevertheless, this edifice combines extreme lightness—reduced to a height of 30 centimeters, the tower would weigh less than 10 grams—with a base that allows it to withstand winds of more than 150 kilometers an hour.

Animalcules that appeared 600 million years ago are also light and resistant thanks to their filigree skeleton. The radiolarians are marine pro-

Figure 2.2
The Crystal Palace of New York. Constructed on the same model as the Crystal Palace of London, the Crystal Palace of New York was also entirely destroyed in a great fire.

tozoans invisible to the naked eye. They are made up of a single cell surrounded by a rigid radial structure composed of silicon and built according to a geometric plan specific to each species. In the course of evolution, the organization of their skeleton was modified by economizing on the constitutive material, which had become rarer in the environment, so the radiolarians of the Quaternary are four times lighter than those of the Eocene. Lighter, but still as resistant.

This combination interested French architect Robert Le Ricolais, professor at the University of Pennsylvania in Philadelphia in the 1950s, who elaborated very avant-garde structures by copying the radiolarians perfectly sketched by the German biologist Ernst Haeckel in the nineteenth century.

Forms

Forms present in nature have mathematical properties that have inspired inventors.

In 1420, appearing before the eminent committee that questioned his competence to construct the cupola of Santa Maria del Fiore in Florence,

Figure 2.3
Left: Robert Le Ricolais in front of one of his architectural structures inspired by radiolarians. Right: Reticulated and radiolarian systems drawn by Ernst Haeckel. (© René Motro, Université de Montpellier II.)

Filippo Brunelleschi rapped an egg on the table and crushed the shell lightly at the base. Everyone could see that the egg remained vertical and immobile, whereas if the architect had tried to explain in words the planned construction, to do it would have seemed impossible. This egg presented in Florentine style (later used by Christopher Columbus) must have been convincing because Brunelleschi obtained carte blanche, and the gigantic dome that he built is still in place. He conceived it (by coincidence?) according to a principle that recalls the eggshell, so that bricks laid fishbone style mutually block each other and thus totally bypass the need for classic supports such as wooden arches. An ingenious encasing of two cupolas likewise ensures a good distribution of pressure.

An egg obviously does not contain arches, which would obstruct the embryo in development, yet its structure is capable of resisting considerable force: the shell of a hen's egg is 0.3 millimeters thick and can withstand up to 3.0 kilograms of pressure, and that of an ostrich egg is ten times thicker and can withstand a pressure twenty times stronger! The shells exploit a "trick": the crystals of the mineral salts that constitute them are oriented toward the center of the egg and are self-blocking, like the bricks in the dome of Florence. Various architects—including the Germans Frei Otto, Carl Zeiss, and Heinz Isler—have recently adopted these characteristics to calculate the exact thickness of the layers needed to cover various domes or to establish force lines for immense stretched structures.

The oblong form is known to move rapidly in the air and in water without too much effort. Bizarrely, marine animals are also the best models for terrestrial or aerial mobiles! For example, the engineers of Daimler AG are currently trying out the form of the tropical boxfish to save precious liters of fuel in future automobiles. This research has resulted in a prototype presented in Washington in 2005. Although the animal presents an almost cubic belly, this characteristic makes it paradoxically very hydrodynamic and even, according to measurements undertaken by engineers, more aerodynamic than current cars, with a coefficient of air penetration of 0.06 instead of 0.30![1] The "Mercedes-Benz bionic car" would boast a coefficient of 0.19 and could save 20 percent in fuel and 80 percent in nitrous oxide emissions. Its body, like the rigid carapace of the fish, is conceived as if it were composed of numerous hexagonal panels supported by a metallic vertebral column. Such a car would thus be very competitive in both lightness and robustness, the two great principles of both nature and the automobile industry.

Even an airplane of the future might be shaped like a fish, too. The Smartfish, being developed by several international companies and research laboratories at the initiative of Swiss engineer Koni Schafroth, is inspired by the form of several fish, especially the tuna, the fastest and most agile animal in the sea. A meterwide miniprototype has already successfully achieved first flight, in April 2007.

 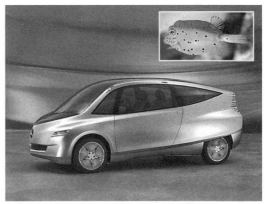

Figure 2.4
Left: The Smartfish plane. (© Koni Schafroth, Team SmartFish GmbH.) Right: The boxfish car and its model. (© Mercedes-Benz Cars, media materials and pictures, Stuttgart.)

Arising from the same inspiration, a very high-performance bicycle, modeled on the form of Antarctic penguins, is being developed and might soon invade the squads of the Tour de France. It so happens that these animals' morphology allows them to spend very little energy as they swim because it reduces to a minimum the water drag that holds them back. According to Rudolf Bannasch and his team at the Technical University of Berlin, if penguins were vehicles, they might travel 1,500 kilometers on only a liter of gasoline! This ability inspires the conception of new forms of land vehicles, including the cycles mentioned earlier as well as proto-types for planes and submarines.

Let us end this catalog of bionic-inspired vehicles by mentioning the Shinkansen 500 series, a Japanese high-speed train whose nose imitates in form the head and beak of the Common Kingfisher, which allow this train to gain speed and economize energy during its passage through tunnels. In effect, the problem that a train has in adjusting to differences in air resistance outside and inside tunnels strongly resembles the problem that the bird solves when it travels easily between media of unequal density, such as air and water.

A Golden Form

What do a sunflower, the Strasbourg cathedral, a snail's shell, Concerto No. 3 by Bartok, a pine cone, the Saint Lazare station as painted by Monet, the heart of a daisy, Le Corbusier's Radiant City, a pineapple, and a Stradi-varius all have in common? The golden number.

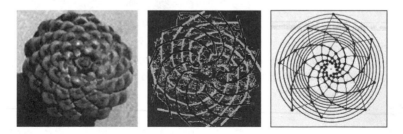

Figure 2.5
A pinecone and the golden number. Left: The scales of the pinecone form spirals whose numbers correspond to the Fibonacci series. If one represents the four corners of the scales by points and then connects all these points, one obtains some spirals that turn to the right and others to the left. Center: There are eight green spirals in one direction, thirteen red spirals in the other direction: 8 and 13 are two consecu-tive terms in the Fibonacci series: 1, 1, 2, 3, 5, 8, 13. Right: Each point belongs to two spirals. The numbers of points on each of these spirals are also two numbers in the Fibonacci series. (© Dunod). See color plate 3.

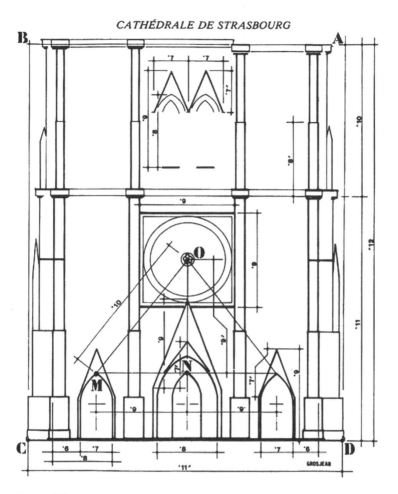

Figure 2.6
The facade of the Strasbourg Cathedral is inscribed in a golden rectangle ABCD, and the many architectural details of this facade are linked to the golden number. (© Dunod.)

We find this golden number in the organization of many natural forms. For example, we notice that ammonites and the snail are constructed on a spiral in which the relation between the successive radiuses corresponds to the golden number. In the same way, the hearts of sunflowers and daisies and the scales of pinecones and pineapples are arranged in several spirals whose successive numbers correspond to two consecutive terms in the Fibonacci series. The numbers of leaves found on a stem until encountering a leaf directly above the starting one also correspond to this sequence of numbers.

The Golden Number and Fractals

Leonardo de Pisa, called Fibonacci, was an Italian mathematician of the thirteenth century who studied a series of numbers in which each successive term is calculated by adding together the two preceding numbers. This calculation was originally supposed to determine how many rabbits might be produced by an initial couple in a year, accepting that each couple would produce a new couple each month and that there were no deaths. Starting from 1 (the second number of the series being 0 + 1 = 1), the series would give: 1, 1, 2, 3, 5, 8, 13, 21, 34, 55, 89, 111, and so on. One of the particularities of this sequence is that the ratios between two consecutive terms (3:2, 5:3, 8:5, etc.) converge on the same result, the *golden number*—called ϕ (Phi) in honor of the sculptor and architect Phidias—which is approximately 1.618. Thus, the golden rectangle is defined as one whose length-to-width ratio is equal to ϕ; the golden triangle is an isosceles triangle whose hypotenuse-to-shortest-side ratio is equal to ϕ, and so on. According to Vitruvius, "proportion consists in taking a fixed module, in each case, both for the parts of a building and for the whole, by which the method of symmetry and proportion is put into practice."[a] It is on this model that Leonardo da Vinci drew the famous "Vitruvian Man."

The principle of the Fibonacci series can be generalized: it is a matter of repeating a motif at different scales. This is the definition of a *fractal*, a term created by the French mathematician Benoit Mandelbrot in the 1970s to refer to any figure that presents a phenomenon of self-similarity at various dimensions. It may involve figures whose motifs repeated at various scales are identical (such as the successive chambers of the nautilus, the leaves of a fern, or the elements of a horsetail) or just vaguely similar (such as the arborescence of the blood vessels and capillaries of our blood system, the ramifications of trees, or the irregularities of oceanic coasts).

The *L-system* presents similarities with the Fibonacci sequence and fractals. It is a mathematical formalism conceived by the Hungarian biologist Aristid Lindenmayer in 1968, with the goal of presenting a general model of the development of living systems. The basic elements are modeled with the aid of symbols. In each generation, the elements are reproduced following simple rules, which may give at the end a very complex form. Such a system can model a fern, a snowflake, or the shell of a gastropod.

Note

a. Vitruvius states here that a harmonious relation links one part (a) to the whole ($a + b$) if one has the ratio $a/b = (a + b)/a$. If b is set at 1, the solution to this equation is $a = 1.618$ (i.e., the golden number).

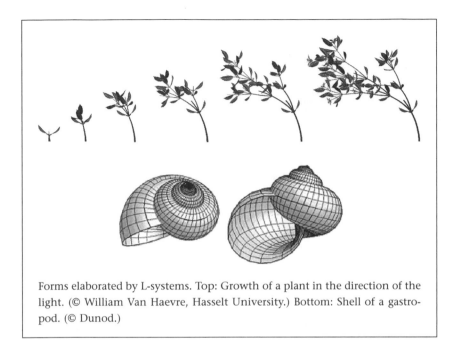

Forms elaborated by L-systems. Top: Growth of a plant in the direction of the light. (© William Van Haevre, Hasselt University.) Bottom: Shell of a gastropod. (© Dunod.)

These "divine proportions" have long been recognized and have been used in the construction of buildings, the composition of paintings, and the composition of musical works. Their best-known applications in architecture are those of the French architect Le Corbusier, who introduced them into several of his works in the twentieth century, whether in their exterior aspect (for the Radiant City in Marseille, especially) or in their interior organization (the arrangement of stained glass windows in the Ronchamp Chapel—whose form was inspired, moreover, by the crab's carapace). He even developed a tool, the Modulor, that enabled the calculation of ideal dimensions, such as those of a room and its furnishings adapted to human proportions.

Architects, including Le Corbusier again but also many other contemporary architects, have exploited the repetition of similar motifs found in so-called fractal forms. They use these forms to try to resolve the problem of lodgings in limited space. The French architect Ionel Schein has created a house with organic growth in the form of a snail and hotel rooms in stackable capsules. Similarly, the Archigram Group is studying a "Plug-In City": a whole adaptable town contained in a building within which the spatial elements assigned to various activities can be moved, fused, or replaced.

The same repetitive logic is at the heart of the work of Dennis Dollens, an architect who lives in New Mexico and teaches at the University of Catalonia. His research aims to design habitations whose structure and

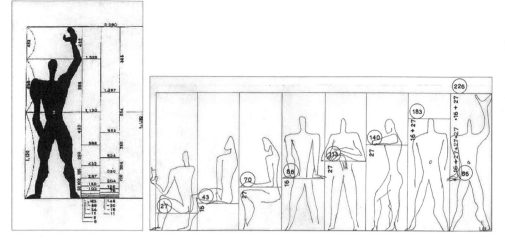

Figure 2.7
Le Corbusier's Modulor and its use to calculate the heights of furnishings adapted
to human sizes (1945, Archive FLC B3-(20)11 (c)FLC/ADAGP, 2007).

functions are inspired by those of plants. By means of adapted software, he
uses Lindenmayer's L-systems to simulate on the computer the growth of
vegetal structures of very complex appearance but based on simple laws of
development. Thus, the Arizona Tower project capitalizes on knowledge
acquired on the morphology and physiology of plants that grow in the
Arizona desert, with a view to constructing an edifice that will be closely
integrated into nature. The development program that served in the tower's
conception has generated its solar panels as if they were leaves, its living
chambers as if they were pods, its roots and load-bearing structure as if it
were a tree. Such an edifice "will survive" by drawing water from the ground,
by storing, recycling, and purifying that water according to its needs, and
by modulating its caloric intake and losses according to the external temperature, detecting and adapting to threats from pollution or allergens.

Other projects, such as "Fab Tree Hab" led by architects and engineers
from the Massachusetts Institute of Technology (MIT), aim in the same
way to make habitations that will grow like elements in a natural ecosys-
tem and that will fulfill all their own functions. From the same perspective,
the design for the future nanobiomedical and biological Chengdu Institute
in China, conceived by Sloan Kulper, is inspired by the morphology and
physiology of a living cell.

Modern builders do not hesitate to take the flame handed on by the
Catalan architect and artist Antonio Gaudí, a prestigious student of nature

Figure 2.8
The Arizona Tower project. (© Dennis Dollens, Universitat Internacional de Catalunya, Barcelona.) See color plate 1.

Figure 2.9
The Fab Tree Hab project. (© Mitchell Joachim, Terreform.) See color plate 2.

who took inspiration from it for various pieces of furniture, art objects, and buildings—including the astonishing Sagrada Familia cathedral in Barcelona, begun in 1884 and never finished. The inventiveness of his work remains unequalled to this day. Yet it was decried in his lifetime and aroused acerbic criticism from the writer George Orwell, who, passing through Barcelona around 1936, was sorry that the anarchists did not seize the occasion of the Civil War to burn all of Gaudi's creations!

Textures

Animals' bodily envelopes are polyvalent. They protect the fragile parts or those subject to predation. They ensure exchanges between the organism and the surrounding environment. They sometimes also fulfill quite specific functions linked to climate or their hosts' very particular way of life—

Figure 2.10
Artist's view of the Chengdu Institute and its interior garden depicting a few cellular
organelles: vacuoles, mitochondria, and so on. (© Sloan Kulper, Kennedy & Violich
Architecture, Ltd.)

all thanks to complicated textures that have nothing in common with the smooth surfaces that seem to represent for us humans the height of manu-factured perfection.

From this standpoint, the examples of the scarab beetle and of some plants such as the eucalyptus, coffee tree, date palm, acacia, and agave—all organisms that one might call "demystifying" or resourceful—are particu-larly edifying.

Certain plants are known for their capacity to gather mist—also called "horizontal precipitations." No doubt the most spectacular is the Garoé or fountain tree, the symbol of the Canaries, which has now disappeared but could "produce" about 80 liters of water per day. All these plants are able to resist very hot climates. Their adaptation to drought is translated into the protection of surfaces of evaporation, which are often covered with a sort of wax and sometimes equipped with tiny filaments that can retain humidity. The capacity to glisten on the surface proceeds not only from their texture, but also from their placement: these plants must be perched sufficiently high to be ventilated and be able to maximize the turbulence of the air around them.

These characteristics have been reproduced in the design of immense nets that capture by condensation the fine droplets of the *camanchaca*, the fog that is formed above mountains in the Atacama Desert of Chile. The first fifty prototype nets, each with a surface area of 48 square meters, were installed in 1987 in the arid region of El Tofo. They were made of 0.1-mil-limeter-thick polypropylene, whose texture allowed the banking of drops, then their sliding by gravity into a gutter situated at the foot. The water was then transported to a reservoir situated a few hundred meters away that served the fishing villages of Chungungo and Los Hornos. A few years ago, seventy-five of these nets could still supply 40 liters of water per person per day. However, the project was gradually abandoned, victim of a success that resulted in a sudden influx of population, but other nets are currently functional in other dry regions, such as Yemen.

The Namibia Desert in southwestern Africa is even more arid than El Tofo. It is undoubtedly the hottest place on earth. The lizards that live there are obliged to dance on the sand to avoid burning their feet. Still, every morning a very dense fog forms, but it cannot develop into rain because the wind disperses it too quickly. Before this happens, though, a strange ritual can be observed: the scarab *Stenocara* prostrates itself facing the wind and seems to meditate the whole time this benevolent mist is passing. In fact, the scarab is drinking: microdrops are periodically formed on the surface of its wing cases and run into its mouth. How can these

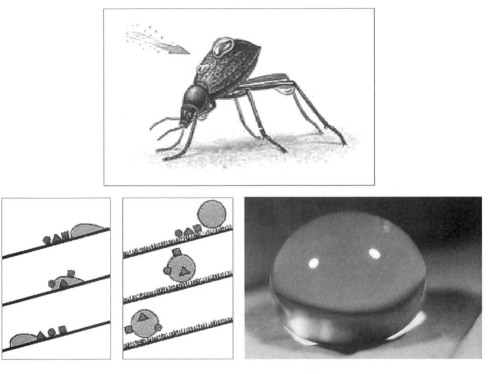

Figure 2.11
Surfaces partially or totally hydrophobic. Top: The scarab *Stenocara*. Bottom left: The lotus effect: on a smooth surface, dust is merely displaced by a water droplet; on a rough surface, the dust is carried off. Bottom right: The interactions between the rough hydrophobic surface and a drop of liquid occur only at a few points, which has the effect of maintaining the spherical form of the drop and provoking its flow down the least inclination of the surface. (© Wilhelm Barthlott, Nees Institut für Biodiversität der Pflanzen, Bonn.)

drops form and drip quickly enough not to evaporate or be carried off by the wind?

In the United Kingdom, zoologist Andrew Parker and engineer Chris Lawrence have deciphered this mystery. Using an electron microscope, they have discovered that on the scarab's rigid wings are constellations of small bumps, whose flat summits are covered with a hydrophilic substance. The slopes of the bumps are punctuated with flat domes that are 10 microns in diameter, arranged in hexagons, and coated with hydrophobic wax. Thus, the molecules of water accumulate at the top of the bumps without evaporating or being carried away by the wind because they are electrostatically attracted to the hydrophilic substance. When a drop attains a certain size, gravity forces it to slide from the top of the bump into the insect's mouth, thanks to hydrophobic slopes.

These researchers—and others since—have tried to fabricate a similar texture. They arranged very small balls on a glass surface and covered them with wax. Then they poured alcohol on the summits of the balls to make them hydrophilic. They noticed that at whatever temperature, a light dew could be collected with efficiency. Even better: by arranging the tiny balls randomly like the scarab's bumps—instead of ordering them regularly—they increased their chance of capturing water, which always flows in an unpredictable way.

In 2006, MIT researchers led by Robert Cohen and Michael Rubner succeeded in perfecting an artificial surface of this kind. Apart from its "demisting" property, this surface might have other applications. It might serve in the elaboration of "microfluid chips"—that is, miniaturized systems able to be introduced into cells in order to perform, in minuscule cavities filled with microfluids, hundreds of different biochemical analyses.

These same teams are also involved in research on the "lotus effect," demonstrated in the 1990s. The symbol of purity in Asia, the lotus is remarkable for its permanent cleanness, even when it grows in swampy or muddy places. It so happens that its texture rejects any molecule of water falling on its surface, and that water sliding over its leaf carries away any other substance found in its passage. Thus, the plant is self-cleaning! A potential godsend for the facades of buildings, for textiles, or for urban housing.

The lotus shares this characteristic with many other plants and some insects. This feature serves principally to evacuate bacteria that might infect them. The principle is not at all the same as with the scarab; the lotus's surface is entirely hydrophobic! Such a surface would not suit the insect, for drops would not have the time to grow and the wind or heat would quickly evacuate them. In examining the surface of these self-cleaning plants and insects in the microscope, one perceives that it is made up of small irregular bumps that prevent a water drop from adhering totally to the surface, as would be the case on a totally flat surface. It is again thanks to these little bumps and to the nature of their interactions with the molecule of water that the latter can carry away any dust as it passes over the plant's or insect's surface.

The fabrication of such cladding arises from nanotechnology, working at a scale that can go to a billionth of a meter. By these procedures, a researcher at the School of Industrial Physics and Chemistry in Paris has achieved a surface of superhydrophobic nanometric spikes. German researchers have also put on the market an aerosol that can be pulverized on any surface whatever. It is made of hydrophobic "nanopowders" mixed with various waxes. After projection, these materials spontaneously organize themselves into microbumps while drying on their support surface. The Lotus-Effect® is a registered trademark for these micro- and nano-structured surfaces, publicized as the only ones that are truly self-cleaning. They are found especially on the optical sensors located at the toll booths of German auto routes.

The bumps have other unexpected properties. Take a smooth object and a striated object and throw them under water with the same force: Which will go faster? In fact, the latter will. The friction of the object in the water will be considerably reduced because the water will be channeled into the object's ridges instead of generating microturbulences!

Few engineers had applied this principle until biologists closely examined the epidermis of sharks. The "teeth of the sea" merit this nickname because their epidermis is covered with tiny scales, or placoids, of the same material as their teeth and arranged in staggered rows. In certain countries, shark skin is used as an abrasive, and in the ocean small fish do not hesitate (despite the danger) to come and rub against the shark to get rid of their parasites, which means the sharks' roughness is sought after! Each placoid is striated and varies in form and size according to the species, sex, and even the place on the individual shark's body. Australian researchers have analyzed minutely the placoids' function and noticed one constant among several species: the placoids of the front of the body have streaks that are reduced in both height and in spacing than those on the rear, which diminishes water turbulence as much as possible and allows sharks to swim rapidly. Another advantage of this reduction in hydrodynamic wake for this predator: it diminishes the noise generated by the movement in water, which seriously complicates the task of the shark's prey.

These placoids have several other functions, which is why they are sometimes called "the Swiss army knife of elasmobranches" (the latter word designating the order to which sharks belong). Placoids serve as protection against potential superpredators, against parasites, and against scraping over rocky seafloors. They are also armed with sensory receptors able to perform instantly a chemical analysis of the importunate being that comes into contact with them.

Engineers (for the time being) have deduced from these denticles' capacities only their power to reduce turbulence and to protect against parasites. The former intensely interested aeronautical engineers of the Cathay Pacific company. They have tested a coating designed to cover a part of the Airbus

Figure 2.12
Placoids of the shark skin seen through an electron microscope. (© Dunod.)

A340 that would allow it to diminish its aerodynamic resistance by 6 percent, thus entailing a saving of about 350 tons of fuel per plane per year. Nevertheless, the weight and cost of maintaining this coating were considered to be too high, so the corresponding tests seem to have been abandoned. However, a similar texture is already used in a Speedo swimsuit that was worn by world-record holders of both sexes at the championships in Shanghai in 2006 and at the Olympic Games of Beijing in 2008.

The second function of the denticles that interests engineers is their antiparasitism. In effect, these structures move lightly in the currents and prevent the anchoring of various animal or vegetal organisms. This is exactly what boats need to prevent their hulls from collecting algae that proliferate and diminish a boat's speed. A covering composed of billions of small, 15-micron spikes has been achieved in the United States. Researchers have even accentuated the movement of these tiny points by a weak electrical current, which compromises even more any chance that sticky grains of pollen will adhere to the boat.

Although rather costly, this project is welcomed by seafarers, for the antialgae covering based on copper that they currently use is toxic and has a dangerous tendency to accumulate in ports.

3 Processes

When one observes nature, one discovers jokes of a superior irony.
—Honoré de Balzac (1837)

The adaptation of structures does not always respond to all the situations with which living systems are confronted. Various processes, simple or sophisticated, also contribute to their survival and can inspire engineers. We choose some in the following categories: how best to adhere, how best to deploy, how best to propagate—and to be propagated.

How Best to Adhere

The official description of the bionic invention that is probably the most celebrated goes as follows: Velcro®—a contraction of the words *velours* (velvet) and *crochet* (hook). We owe this innovation to the Swiss engineer Georges de Mestral, who, coming home from a walk, noticed that burdock flowers remained stuck to his dogs' coats and to his clothes, and it was very difficult to get these prickly burrs off.

He observed in the microscope the way this plant gripped various supports, and he had the idea of fabricating out of nylon small hooks similar to those the flower possesses. The latter are oriented in all directions, which maximizes hitching on to any support, and they are flexible, which facilitates (to some extent) the unhitching. After several improvements, he deposited a patent in 1955. Since then, this brand has spread throughout the world; several million kilometers of it are sold every year. Velcro® attaches parts of clothing and shoes and is used in aviation, automobiles, sports, space, and hospitals. It has also found a playful application in which children or adults can stick to walls, defying the laws of gravity.

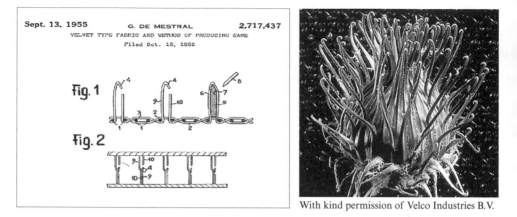

Figure 3.1
Left: Extract from the Velcro® patent. Right: A burdock flower. (© Dunod.)

. . . But Not Too Much!

Velcro® is certainly convenient, except when one tries to attach and detach the two supports in a hurry, an exploit that climbing insects and the gecko lizard perform effortlessly when they move over vertical surfaces.

The German engineer Stanislav Gorb has studied the legs of the fly. At these extremities are two protuberances—pulvilli—that are covered with microfilaments—setulae—whose ends are spatulate. Millions of setulae produce a sweet and oily substance that offers sufficient adherence to the support.

The fly has several means of getting rid of this adherent substance at each step: either it performs a minirotation of its foot, or, if that does not suffice, it removes the substance with the two claws of another foot. This process, based on so-called humid adhesion, is very effective.

The fly, like other climbing animals, possesses another solution for countering heaviness: the van der Waals force, a dry adhesion that works without a sticky substance and that is repositionable at will. But the fly seldom uses this solution due to the efficiency of the preceding method. In contrast, the gecko lizard does use this force exclusively, as demonstrated by Kellar Autumn's team in Portland, Oregon. This lizard has a spectacular capacity for adhesion: it can suspend itself on any surface, smooth or rough, humid or dry, by supporting its whole weight with a single digit. It can also adhere and "disadhere" just as easily up to fifteen times per second, with each operation lasting about 1/8,000 of a second.

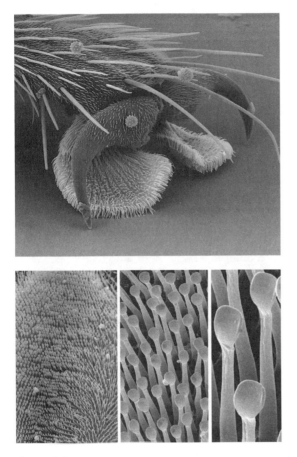

Figure 3.2
Anatomical details of a fly's leg. Top: The claws and pulvilli at the end of a leg. Bottom: The pulvilli are covered with setulae ending in spatulates. (© S. Gorb, S. Niederegger, J. Berger, Max Planck Society, Germany.) See color plate 4.

On each of the five digits of a gecko's foot are found in serried ranks millions of setulae similar to those of flies, but each setula ends in about one thousand tiny spatulates, each half a micron in diameter. A setula has a precise orientation that lends itself to any support in order to maximize the contact of each nanospatula in such a way as to generate the famous van der Waals attractive force. Very weak at the scale of a single atom, this force becomes colossal when it issues from very numerous elements: thus, from all of a gecko's setulae, a weight of about 130 kilograms can be suspended!

The Van der Waals Force

These interactions bear the name of the Dutch physicist Johannes Diderik van der Waals, the Nobel Prize winner in physics in 1910. They are very weak forces that can be understood only in the framework of quantum physics, where they are conceived as the result of the exchange of virtual photons between atoms. As such, they have nothing to do with phenomena occurring at a larger scale: they do not concern, as in classical chemical interactions, the exchange or sharing of electrons. We describe them here in an evidently simplified way.

These interactions involve atoms or molecules characterized by electromagnetic dipoles—that is to say, having opposite charges separated by a certain distance, a little like the two poles of Earth. The negative electrons that gravitate around the atom's positive nucleus entrain at this nanoscopic level a change of polarity according to whether they are grouped at one certain place or another of this atom. This polarity change generates a sort of vibration, characterized by the frequency of the changes. When the vibrations generated by different atoms or molecules are synchronous, then there is an attraction, created by electromagnetic forces, between these atoms or molecules. When the vibrations are asynchronous, then there is repulsion. These interactions can be permanent or temporary. They are of very weak intensity—one hundred times weaker than a classic covalent chemical interaction—but always present. They are what, along with hydrogen bonds, maintain the cohesion of any living material!

The most astonishing thing is that the lizard detaches itself just as rapidly as it attaches itself to the support. When it puts its foot down, it makes its setulae rigid in order to increase each spatula's surface of adhesion; when it lifts its foot, it folds its now-supple setulae so that the spatulates make an angle of more than 30 degrees with the support, which breaks the van der Waals force! In fact, as a consequence of similar principles, it is easier for us to detach an adhesive ribbon by pulling it perpendicularly rather than pulling it in parallel with the surface that supports it. According to these principles, any surface, whatever its chemical composition, is capable of presenting analogous properties, provided that it is divided into nanoscopic elements that are reproduced in very great numbers and that are instantly foldable. Let us add that the spatulates are hydrophobic—like the leaf of the lotus described earlier. Therefore, the dry and infinitely repositionable superadhesive surface is also self-cleaning!

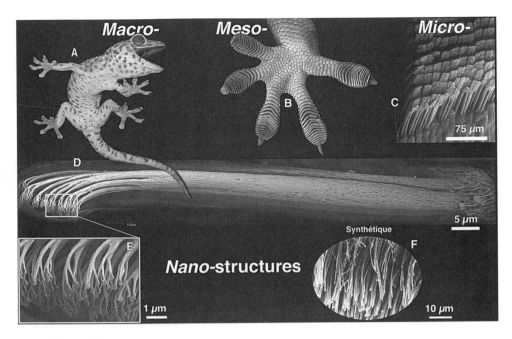

Figure 3.3
A: Gecko. B: One of its feet. C: Setulae arranged on each digit. D, E: Spatulates at the end of each setula. F: Synthetic realization. (© Kellar Autumn, Lewis & Clark College, Portland.) See color plate 5.

Robert Full and Kellar Autumn, at the University of California at Berkeley and Lewis and Clark College in Portland,[1] in collaboration with other American teams, are developing bands made of polypropylene fibers, each 20 microns long and 0.6 microns in diameter (i.e., one hundred times finer than human hair) in groups of 42 million per square centimeter. The capacity of adhesion is made by the changes in the geometric and mechanical properties that the nanofibers undergo as soon as they approach a support. For the time being, the material's efficacy has been demonstrated only in a gecko toy and a robot—as we will see later. Researchers have indeed considered the possibility of producing enough of this covering to suspend a student from one of the windows of the highest building on campus, with a view to ensuring certain publicity for their work, but they gave up the idea—maybe due to a lack of volunteers!

Envisaged future applications of this type of process include covering the soles of athletes' shoes so they can improve their performances during rain and the soles of astronaut's footwear so they can stroll on the walls of spaceships during inspections or repairs.

Yet there remains a final and knotty problem to resolve: if two adhesive bands are too close together, they will adhere to each other. However, in the gecko species' memory, there has never been a case of one gecko's foot stuck to another's. So the secret of dry adhesion is not completely resolved![2]

How Best to Deploy

. . . To Develop

In nature, a seed, a bud, or a chrysalis practices the art of folding in order to unfold leafs, petals, or wings. The ladybug uses it daily to protect its fine membranous wings under its two rigid wing cases. Fragile large surfaces thereby remain intact, despite the tensions occasioned by rapid development or deployment. Humans probably began to use folding about three thousand years ago, the approximate date of the first folded map known in the world.

In 1970, Japanese professor Koryo Miura of the Institute of Aeronautic and Space Sciences was inspired by natural processes to study mathematically how to realize "rigid" origami, where the surfaces between the folds do not bend during folding. A sheet of paper prepared in this way passes from a minimal to a maximal size—and vice versa—by simply pulling two opposite corners, and this gesture can be applied many times without the folds suffering the smallest tear. Rather than classically folding the sheet according to orthogonal folds, Miura invented a process inspired by the hornbeam leaf, which organizes the folds so that some form sharp angles with others and so that any action on one fold has an effect on the others. This process ensures the resistance of sides and edges, and it allows particularly easy refolding—which any user of a roadmap desires.

Many realizations of this process, from the simplest to the most sophisticated, have been patented under the name "Miura-ori folds" and have resulted in many applications by labs or companies throughout the world. Let us mention the folding of solar sails, those huge reflective surfaces that are pushed naturally by photons and that avoid becoming too heavy by carrying an excess of the fuel needed to orbit around the sun. They can attain 20 meters in span by being about a hundred times finer than a sheet of paper, which obviously prevents them from being deployed at takeoff. Among other applications involving this type of folding, let us mention a portable telescope that expands like a flower without any motor, conceived at Lawrence Livermore National Laboratory in California, or the immense retractable roofs whose feasibility was studied by a lab in Cambridge, England.

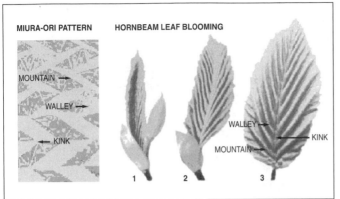 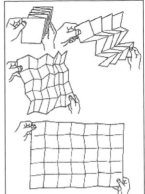

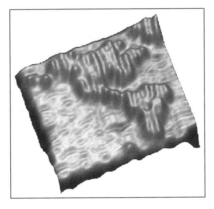

Figure 3.4
The art of origami. Top: The development of a hornbeam leaf generates Miura-ori-type folds. Folding and unfolding of a sheet of paper can take place without changing the placement of the hands. (© Lakshminarayanan Mahadevan, Harvard University.) Bottom: A map of the Americas in molecular origami. (© Paul Rothemund, California Institute of Technology.)

An even more audacious project concerns a car that would fold in case it encounters a pedestrian, thus absorbing the energy generated during a hard shock; in the same spirit, a Tokyo laboratory has designed a whole range of furniture to resist earthquakes.

Even more curious is the case of Paul Rothemund, a California researcher who gives a recipe for designs realized with molecular origami: for example, ad hoc folding at the level of interactions of some DNA chains can form reliefs to represent a map of America. Such designs are visible only by means of tunnel scanning or atomic force microscopes, and they open up unprecedented perspectives in cryptography.

Figure 3.5

Wings and scales of Morphos. Left: Changes in color of a butterfly in the Morphos family. (Picture from Serge Berthier's *Iridescence, les couleurs physiques des insectes* [Paris: Springer France, 2003], with kind permission of Springer Science and Business Media.) Right: Scales of the Lepidoptera in a scanning electron microscope. (© Joseph Le Lannic—CMEBA—Université Rennes 1). See color plate 7.

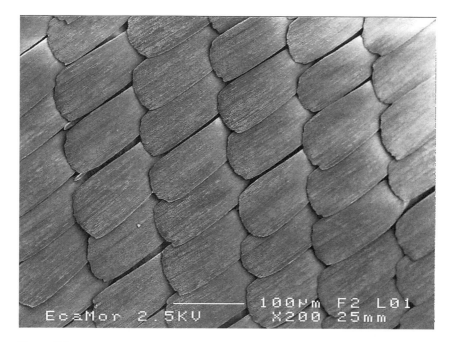

Figure 3.5
(continued)

. . . To Change Color

Butterflies, belonging to the order of Lepidoptera, named so because they
have scaly wings (from the Greek *lepidos*, "scales"), deploy their wings at
the moment of metamorphosis by the same process described previously.
Despite the total absence of colored pigments, the wings of certain species
present brilliant colors due to the fine structures of their scales, which
engender interference and diffraction of the light, grouped under the name
iridescence.

Thus, South American butterflies of the Morphos family present wings
ranging in color from brown to a splendid electric blue, passing through
an azure blue, thanks only to incident light. Still, if the bright blue attracts
partners, it also attracts predators. To remedy this disadvantage, the insect's
rapid wing beatings and the disordered trajectories of its flights make it
visible only during very short periods and at scarcely foreseeable points,
which effectively complicates the predators' task.

A great specialist in the subject, Serge Berthier[3] of the Nanosciences
Institute at the University of Paris 7, has meticulously studied this iridescence process. A San Francisco company applied it to conceive the most
recent generation of flat screens: iMoD (from the term *interferometric*

modulator), made of a reflective metallic membrane covered with nanomirrors—eighty thousand per square centimeter. When electric tension is applied between the membrane and the mirrors, an electrostatic force inclines the mirrors to three different positions, respectively reflecting the three primary colors (red, blue, green). When the mirrors are not inclined, they reflect back black. This type of screen allows a precision of colors and a resolution of the image superior to those of classic screens. Moreover, it consumes a tenth of the energy necessary for a liquid crystal screen because the mirrors use the ambient light and not the artificial white light dispensed at the background. It is for this reason that this technology is widely exploited in portable telephones, for the classic screens consume about 30 percent of the total energy used by the apparatus.

How Best to Propagate: Propagating Liquid or Gas Products

. . . And Dive Like a Nautilus
Because life was originally born in the ocean, it is evident that some living systems would inspire processes that allow better propagation in the liquid element. In particular, many cephalopods' capacities for acceleration have interested engineers. The nautilus, for example, very abundant in the Mesozoic period but not represented much today, hunts in the great depths of the Pacific and Indian oceans. As it grows, it constructs and inhabits the final stage of its shell, always more spacious than the preceding stage. The empty chambers are filled with a mixture of water and gas with the aid of a tube—a siphon—that traverses all the developmental stages. To move, the nautilus suddenly propels gas through the chambers. Thus, water is expelled, provoking a movement in reaction that costs little energy and is very effective. In order to dive to different depths—up to 400 meters—and to rise again, the nautilus modifies the proportion of the water/gas mix in the compartments.

It is by exploiting this process that the Cousteau team developed their diving saucers in their world of silence.

. . . Or Aim Like a Bombing Scarab Beetle
The scarab is remarkable: it possesses an uncommon defense system that allows it to project at three hundred times a second a very irritating liquid carried to a temperature of 100 degrees Celsius. This rapid-fire squirting is all the more surprising because the liquid involved comes from a mixture of two products secreted by different glands, so it must be mixed at the last instant. This operation is performed with a little water in a veritable

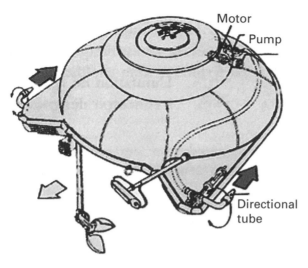

Figure 3.6
Jacques Cousteau's diving saucer. (© Dunod.)

Figure 3.7
Jet of corrosive liquid propelled by the bombing scarab beetle. (© Dunod.)

combustion chamber of only one cubic millimeter. The aim is very precise, too, for the animal is capable of adjusting it with the aid of small abdominal excrescences produced at will.

Tom Eisn, a biologist from Cornell University, is currently working with a professor of thermodynamics and combustion theory at Leeds University, Andy McIntosh, to copy this process of combustion-propulsion. They are trying to resolve the problem of getting gas into airplane turbine engines, a feed that can abruptly stop—especially in cases of extremely low temperatures. The first conclusions indicate that the scarab form of the combustion chamber is of great importance in maximizing the quantity of liquid projected at such high frequency. The form of the jet is also essential to the adjustment of the aim.

Propagating Sounds

. . . And Communicating Like Dolphins

We mentioned earlier that the invention of sonar owes nothing to nature. Nevertheless, the way in which animals produce or exploit sound waves is the source of many other applications. A good reason is that these "mechanical" waves are indeed more adequate than electromagnetic or optical waves for transmitting information in aquatic environments, within which the latter waves are very weakened. It is also not surprising that listening to "conversations" between dolphins inspired Rudolf Bannash and Konstantin Kebkal, two German engineers of the Ministry of Education and Research, to refine an effective system of underwater sound communications.

Dolphins emit two sorts of signals, "clicks" for echolocation to detect obstacles and to orient themselves, and more varied signals such as whistles, groans, and squeaks in order to communicate with each other. The latter signals have been studied in particular. Researchers have observed that dolphins, to understand each other distinctly and to recognize which one of several individuals is communicating, modify the range of the frequency in which they emit their whistles. Thus, the signals coming from several individuals can be perfectly distinguished from each other. On this principle, a first prototype of an underwater model has been refined, allowing the sending of 5 kilobytes of distinct sound messages per second over a distance of 3.5 kilometers.

. . . Or Capturing Them Like an Owl

The famous "small sound" in a car whose source one cannot locate is annoying to many drivers. Taking inspiration from barn and other owls

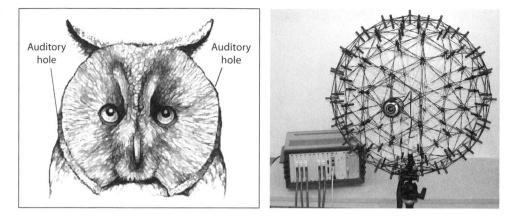

Figure 3.8
Listening like a raptor. Left: Facial mask and asymmetric position of auditory holes among certain nocturnal raptors. (© Dunod.) Right: Spherical "acoustic camera" inspired by these raptors' auditory system. (© GFaI Tech, Berlin.)

might solve this difficulty—with many other undoubtedly important applications. These night birds locate their prey thanks to a visual system exhibiting a light sensitivity that is one hundred times superior to that of humans. And their auditory system allows them even better performances: in fact, they are capable of hunting in the blackest night and even of detecting the muffled sound of small prey moving under the snow.

Their auditory canal is found behind the "facial disk," that characteristic plumed mask that surrounds their eyes. It strengthens the sound waves by directing them precisely to these conduits, with a gain that can amount to 10 decibels. This hypersensitivity permits these birds to measure precisely the distance to where the prey lies. Locating the prey is an even more sophisticated process: a gap in time of 0.0002 second that separates the arrival of the sound at the level of each ear is sufficient for the animal to orient its head. The animal's perception of this gap is affected by anatomical particularities observed in certain species—such as the asymmetric folds of the internal and external ear and the different heights of the left and right auditory holes. Moreover, small and repeated lateral movements of the head refine the detection of this gap.

The Berlin company GFaI has invented and commercialized acoustic cameras inspired by these raptors' auditory system. The cameras are equipped with many microphones—up to 120—that capture sounds emitted at distances ranging from 30 centimeters to 300 meters. A computer processes these sounds and superimposes a visual image of the target on its "sound image." This principle can be applied to know where the noise of a machine can be reduced, to verify good sound propagation, or else to detect the origin of breakdowns.

How Best to Propagate

. . . In the Air, to Accelerate Like a Dragonfly

Engineers have put a great deal of time into understanding why chaser pilots of the Second World War had visual trouble, even blackouts, just at the moment when they should have been particularly vigilant, notably after acceleration in the pursuit of an enemy plane. In fact, the pressure exerted on the pilot might be such that it makes his blood suddenly rush back toward his lower body, depriving his brain of oxygen. Although flying suits invented to trap cushions of air and thus to direct the blood flow in a few seconds were effective in the 1960s, they would be totally useless for current pilots, who have only a half-second to resist an acceleration of 10 g's in chase planes—which submits all their organs to a force equivalent to ten times their own weight.

Swiss and German engineers conceived the "dragonfly suit" after having studied how this insect proves capable of resisting accelerations of 30 G's. In effect, the dragonfly's organs have the particularity of floating in its blood, which circulates without either veins or arteries. The dragonfly's heart is tubular in form and serves to agitate constantly the liquid that

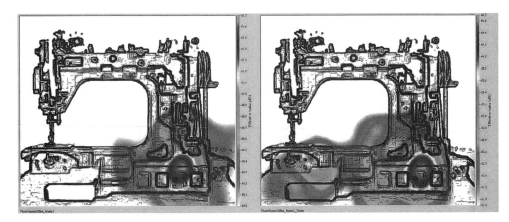

Figure 3.9
"Sound" images of a sewing machine. (© GFaI Tech, Berlin; Axilane Instruments, Nozay.) Left: Operating state. Right: With a spooling problem.

Figure 3.10
The argyronete's elegant diving helmet. (© Dunod.)

surrounds it and the other organs and that protects the whole body against lightning changes of speed. On this principle, a flying suit filled with water (Libelle G-Multiplus) has been created painfully, for designers had to find suitable materials—and find a way to avoid circulating the contents of a bathtub around the pilot's body! This problem was resolved by placing, in a jumpsuit that will not lose its shape, thin bands that go from the shoulders to the heels and are filled with only two liters of water. In case of significant acceleration, the water is displaced toward the legs, exercising a countereffect of pressure in all parts of the body and sending the blood to the brain without any delay. Engineers have even foreseen that this protection and this reserve of water might help in the pilot's survival if he were ejected from the cockpit.

. . . Or in the Water, to Remain Dry Like a Spider

More exactly, remaining dry like a water spider called an argyronete,[4] which inspired a German researcher, Zdenek Cerman, from the same laboratory as the discoverer of the Lotus effect. Entomologist Henri Fabre describes this spider poetically: "The argyronete makes for itself, with silk, within the water, an elegant diving helmet where it stores air. Thus provided with the breathing element, it watches in the cool for the arrival of prey."[5]

Exceptional among spiders, the argyronete leads an entirely aquatic life. Yet its lungs do not allow it to breathe the oxygen dissolved in the water of the pools it inhabits. It must therefore weave a diving helmet in silk that it fills with air in the course of numerous trips to the surface. To do so, it breaches with the head down and uses its feet to imprison microbubbles of air retained by the millions of hairs that cover its body. Each time it surfaces, it deposits a little air in its habitat and then finishes its expeditions entirely dry, always surrounded by its aerated cloak. It is this particularity that interests the researcher. A collaboration with the Denkendorf Institute has enabled the design of a fabric capable of imprisoning ambient air passively—hence without expense of energy—and of remaining dry after a stay of four days in water, ten times longer than any other known textile. The applications are numerous, going from the fabrication of swimsuits that will prevent hypothermia to a covering for boats or pipelines that will reduce the friction of circulating liquids.

4   Materials

The world is a fine book, but of little use to someone who cannot read it.
—Carlo Goldoni (1757)

We deal here with just a few of the most representative materials in current research.

Yet Another Adhesive Product!

Processes of humid and dry adhesion transposed from those utilized by insects and geckos have already been described. The Holy Grail quest for adhesives goes back to the Neolithic Age, when natural bitumen was used to attach points of bones to the stems of arrows. About 3,500 years ago, Egyptians wrote down recipes for vegetal and animal glues. In the fifteenth century, the Western invention of printing and of the bindings associated with it increased the exploitation of naturally occurring sticky substances such as the sap of the acacia or egg whites. After that, the development of the book industry required a diversification of these resources and then the development of synthetic adhesives. Engineers are still looking for a product that can stick to any support whatever and in any environment, aerial or aqueous, that can also be positioned and repositioned instantaneously, and that would be biodegradable without releasing toxic by-products.

. . . Like the Green Lacewing
The green lacewing (chrysope) with golden eyes, a little-known insect, has fabricated for hundreds of millions of years an instant superglue in an aerial environment and without the disadvantages of toxicity. The female lays up to a thousand eggs per reproductive cycle, which she lays delicately alongside each other on plant stems or leaves. Each egg is perched at the

end of a pedicel about one centimeter long, which attaches it to its support and aspirates humidity to it. This pedicel is riveted to its base by a drop of liquid glue that immediately solidifies upon contact with air. Yet, for the time being, the secret of this effective and "clean" product has not been decrypted.

. . . Or Like the Blue Mussel?

In contrast, the secret of a glue that hardens under water and that can resist strong turbulence is almost entirely deciphered. It comes from *Mytilus edulis*, the blue mussel, which produces an adhesive that is strong enough to attach the mussel by the byssus—filaments that come out of its shell—to any support in salty water. This is what interests surgeons, ophthalmologists, and dentists, who with such a product might repair organic lesions in a living body, an aqueous and salty environment like that of the sea.

Various proteins take part in the composition of this product. One protein prepares the support; another is the glue itself; other proteins form the string of byssus; and yet another protein protects the string from bacterial decomposition. Swedish researchers have indeed purified these proteins—ten thousand mussels furnish only one gram of the product—but in doing so they encountered a great difficulty: it was impossible to unstick the product from the instruments that extracted it!

A researcher at Forest College of the University of Oregon preferred to tackle the problem from another angle. Having determined that the gluing properties of the byssus proteins were due to a great quantity of amino acids that present the same groupings in phenol hydroxyl, Kaichang Li had the idea of incorporating this type of molecule in . . . tofu, the protein derived from soy, which is abundant and cheap in the United States. Thus, Li elaborated a very competitive adhesive, particularly as compared to resins of the "urea-formaldehyde" type, which although in common use, are suspected of being irritants as well as carcinogenic to humans. For the time being, the material is adapted only for wood, but it makes plywood much more resistant to humidity, even during prolonged stays in waters that are turbulent or heated to a high temperature.

An Elastic Product

. . . More Resistant Than Steel

In 1709, both fingerless gloves and stockings made from spider silk were presented to the Montpellier Academy by Francois Xavier Bon de Saint

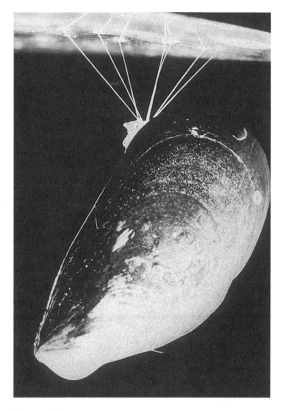

Figure 4.1
Blue mussel and its byssus. (© Herbert Waite, University of California, Santa Barbara.)

Hilaire, first president "en Survivance de la Cour des Comptes Aydes et Finances de Languedoc," who wanted to promote the industrial production of these products.[1] However, the physician and naturalist René-Antoine Ferchault de Réaumur rapidly removed Saint Hilaire's illusions: it would take the silk-producing glands of 55,296 spiders to produce 500 grams of silk, whereas only 2,500 silk worms suffice for this same task; plus, it would also be necessary to trap billions of flies to feed the little workers! Nevertheless, clothes made from this material were woven in Madagascar, where at the end of the nineteenth century a missionary set up a school of artisanal weaving for which Halabe spiders furnished the thread. Robes from this workshop were worthy of the finest raiment, but this material is not only aesthetic. The bionic fiber that constitutes it is lightweight and supple, forty times finer than a human hair, four times more resistant than steel, and more elastic than the synthetic fiber Kevlar®

from which bulletproof jackets are made. A cable of it only as thick as a thumb might hold aloft ten buses and stop a plane in flight, a feat that cannot be performed by the other natural silk, the one produced by the Bombyx caterpillar. Moreover, the spider's thread has a "shape memory" that means it can recover its initial configuration without external stimulation. This property, much studied by Rennes University researchers in France, would explain why the spider, when suspended from its thread, remains perfectly immovable instead of turning on itself as a climber at the end of a rope.

Such qualities would be particularly useful for fabricating fishing nets, mountain-climbing ropes, fiber optics, and the suture thread for ophthalmologic and cardiac surgery—and even for making artificial tissues to replace organic tissues.

Fossil discoveries have shown that spiders have woven this material for 380 million years. They possess from six to nine glands specialized in each sort of thread: several for the web, one to wrap prey, one for cocoons, and so on. According to researchers at the Federal Polytechnic School of Zurich, each thread is made up of a bundle of fibrilla, each being formed of many chains of proteins arranged in alternating zones of layers that are strongly ordered and rigid (giving it robustness) and zones of barely structured chains (responsible for its elasticity). Nevertheless, the secret of this thread resides as much in its structure as in the way it is elaborated by the animal, which alternately produces one filamentous protein and then another protein that regulates the hardening of the former one in the air. This thread-making process is not yet completely understood.

To obtain this precious tiny cord, the most natural solution—raising a battery of spiders—has proved impossible for many reasons, ranging from these animals' hostility to each other to the cost of installations that are large enough to accommodate this character trait. So an artificial solution has been actively sought. In one project, the Canadian company Nexia Biotechnologies, for example, raises transgenic goats to produce the identified proteins in their milk. These proteins are purified and passed under pressure through a microscopic sieve so that they form a bundle of fibrilla similar to the desired thread. However, this new fiber, named BioSteel®, still presents too many problems for it to be commercially exploited. David Kaplan at Tufts University is trying to get around the difficulty of special thread making by allying the products of two cloned genes, one from the spider that produces a fibrous and supple protein, and the other from diatomic algae to furnish the silicon that will ensure its rigidity.

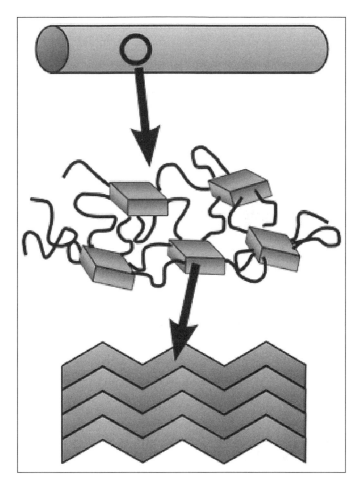

Figure 4.2
Top: A thread of spider silk made of rigid layers (represented by small parallelepi-peds) that are separated by less-structured regions (represented by black filaments in the middle), which give it flexibility and elasticity. Bottom: Rigid layers are themselves structured in a way to give robustness to the whole. (© Dunod.)

Apart from very costly transgenic solutions, research of another type is being pursued, particularly at MIT, where a team under Paula Hammond is not trying to produce a similar substance, but rather to synthesize the polymers of quite another kind than those of the silk spider, but that are able to imitate the properties—both rigid and elastic—of this enigmatic material.

Nanosprings

Imagine a human on the ground floor of a skyscraper who leaps in a single bound up to the balcony on the one hundredth floor. It is what he would manage to do naturally if his tendons were made of *resiline*, the substance that enables a flea to make jumps of about 150 times its size or a fly to beat its wings about 500 million times in the course of its life.

This ultraelastic product owes these feats to the fact that it can restore 99 percent of the mechanical energy that is applied to it, whereas the elastine present in our tendons or polybutadiene—the best synthetic rubber in the world—can restore only 90 percent and 80 percent, respectively. In contrast to these latter products, resiline maintains for a very long time its properties of resilience (i.e., its capacity to regain its original form after having been deformed), a quality necessary for these insects because they never renew the stock of resiline that they acquire at the beginning of their development.

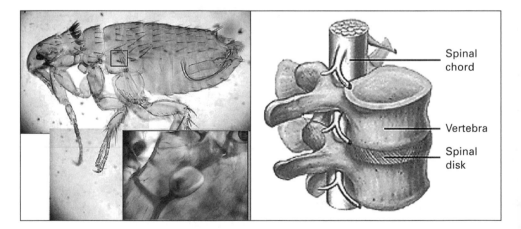

Figure 4.3
Left: A drop of resiline at the level of leg articulations allows the flea to make prodigious jumps. (© Dave Merritt, University of Queensland. Brisbane.) Right: Placement of spinal disks between each of our vertebrae. (© Dunod.)

This protein is constituted in large part by an amino acid, proline, which has the particularity of being "angled" and therefore serving to fold the molecule like a spring. It is impossible to gather enough of the protein from insects. Nevertheless, researchers at the Commonwealth Scientific and Industrial Research Organization of Queensland University and at the National Australian University have been able to obtain a solution of these molecules on the basis of bacteria provided by the cloned gene of the vinegar fly. Their most difficult task was to refine chemical reactions necessary to make the scattered proteins associate with each other and form a material that can be worked in various ways. The ribbon of material thus produced can support an elongation of 300 percent before tearing and can do so many times without its initial form being altered.

The applications of this material include repairing blood vessels and heart valves, but especially replacing our spinal disks, structures that cushion the movements of our vertebrae—unfortunately less resilient than the insect's nanosprings—each time we bend our back, which we do much less frequently than the flea jumps in its life.[2]

Resistors

Best Jaws on Earth
Wood, concrete, metal—nothing can resist them. The materials that rat incisors are made of would be an alternative to the blades of industrial grinders, which are indispensable for the manufacture of granules, the basis of any plastic object. The blades currently get blunt in a few hours, and so grinders are often interrupted for necessary sharpening. These minutes lost each day diminish productivity significantly.

Contrary to most mammals' incisors, the rat's incisors are not entirely surrounded with enamel. This material covers their anterior layer but not the softest ivory that stabilizes the tooth at the back. Because its incisors grow constantly, the rat limits their length by wearing them down; to do so, it continually rubs the enameled surfaces against the ivory surfaces. The tender part gets blunt, leaving in front a particularly sharp edge that is constantly sharpening itself. The enamel of this edge is harder than various metals such as lead, copper, and iron—hence, this rodent's astonishing destructive power.

Marcus Rechberger and Jürgen Bertling, Swiss researchers of the Fraunhofer Institute for Environmental, Security, and Energy Technologies, have been recompensed for taking the rat incisor as a model. With the help of an industrial partner, they have developed a grinder with

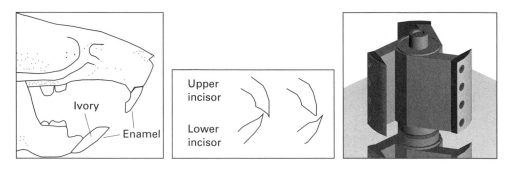

Figure 4.4
Incisors and grinder. Left: A rat's incisors. (© Dunod.) Center: Rubbing the incisors of the upper and lower jaws against each other sharpens them permanently. (© Dunod.) Right: A self-sharpening grinder inspired by this process. (© Marcus Rechberger, Fraunhofer Institute for Environmental, Safety, and Energy Technology.)

autosharpening blades. Convex in form, each blade is covered at the back by a carbide alloy of tungsten and cobalt and in front by several layers of particularly resistant ceramic. As the blades are used, they get sharpened and become finer and finer. Therefore, they grind more and more effectively, using less and less energy to accomplish this work. Proud of their transposition of the "best jaws on earth," the researchers advertise this material with a simple motto: "Install and forget." Maybe they have themselves forgotten that they have not yet invented blades that grow back on their own!

The Ear of the Sea
The gastropod *Haliotis rufescens*, better known as the red abalone or "red ear shell," lives near the coast of California. The interior of its shell is carpeted with mother-of-pearl (nacre) that consolidates its habitat and is continually produced for the purpose of development or repair. This nacreous material, harder than ceramic, is very coveted for making jewelry, which unfortunately will probably be responsible for the extinction of the species. In the course of evolution, the abalone has acquired resistance to predation by being constantly attacked by its principal predator, the otter, which works at breaking the shell with the aid of a stone aimed at the belly.

The abalone makes its nacreous substances in tiles of calcium carbonate—aragonite—that is a thousand times finer than human hair. This material is produced with seawater, a substrate that favors the material's

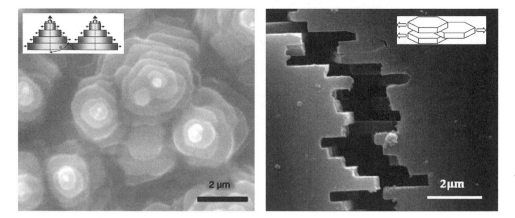

Figure 4.5
Microstructure of the abalone nacre. Left: Hexagonal tiles of aragonite are piled on each other and linked by an adhesive protein, conchyoline. Right: Tiles can slide laterally under the effect of a shock. (© Marc Meyers, University of California, San Diego.)

crystallization. In one layer, each tile is a little offset from its neighbors. Thousands of successive strata are deposited one on the other and are joined by an adhesive protein, conchyoline, which firmly maintains the various layers in relation to each other, except for the tiles in the same layer, which can slide lightly in relation to each other in order to absorb the energy that threatens the shell's integrity. This particularity is the recipe for this material's great resistance to shock.

Several laboratories, including one supervised by Kenneth Vecchio at the University of California in San Diego, have attempted the fabrication of materials inspired by the abalone's mother-of-pearl. The laboratories have perfected a metallic composite composed of superimposed layers of aluminum (replacing aragonite) and titanium (replacing conchyoline). In order to test the stiffness of this product, a shaft of tungsten heated red hot was plunged into a sample 2 centimeters thick at a speed of 900 meters per second. The shaft penetrated only one little centimeter. Extremely promising applications of this research include use in military clothing or for replacing beryllium, a strong but toxic metal commonly used in aerospace applications.

II Behaviors

5 From Automata to Animats

The movement of a gnat in itself contains more marvels than all the art of men can represent; so that if one could buy the sight of all the springs that are in this tiny animal, or else learn the art of making automata or machines that have as many movements, then all that the world has ever produced in fruits, gold, and silver, would not amount to the fair price for the simple sight of such springs.

—Marin Mersenne (1636)

Philosophical Toys

Many legends are associated with the first machines that imitated the behavior of living systems. The reality of certain automata claimed by history appears in fact doubtful, like that of the poet Virgil's bronze fly, which was supposed to have served to fend off all the flies invading Naples. One might also doubt that Gerbert d'Aurillac—who, under the name Sylvester II, was pope in the year 1000—had a talking head that answered "yes" or "no" to any question put to it. Just as suspect is the information that the German mathematician and astronomer Regiomontanus fabricated a metallic eagle that flew before the emperor Maximilian when he entered Nuremberg in 1470. Finally, let us mention a story too good to be true: the humanoid that the philosopher René Descartes built in effigy of his dead daughter Francine. He is said to have taken it in a coffer with him on a sea voyage, but when the ship's captain suddenly discovered this automaton gesticulating, he quickly gave the order to throw it overboard, thinking it the work of Satan.

Those automata of whose existence we are reasonably certain can be classified into three categories: those that tested techniques of their day, those that wanted to be simply ludic, and those that helped us understand better the anatomy and physiology of the living.

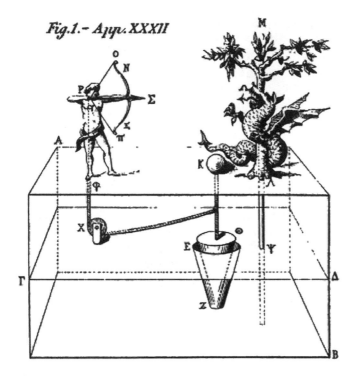

Figure 5.1
Automata of Hero of Alexandria: "On a pedestal is placed a small tree around which
a dragon rolls; a figure of Hercules follows, pulling a bowstring; finally an apple is
placed on the pedestal. If someone's hand happens to lift this apple, Hercules will
shoot his arrow at the dragon and the dragon will hiss." (Hero of Alexandria, *The
Pneumatics*, Bennett Woodcroft, trans. and ed. [London: Taylor and Maberly, 1851],
section 40.)

The representatives of the first category were experiments involving the
effects of hydraulic, steam, or compressed air pressure—such as the wooden
dove of Archytas of Tarentum (fourth century BCE), of which an imprecise
description has come down to us,[1] or the mythic animals and personages
conceived by Hero of Alexandria (first century CE), whose works are relatively well known to us.

The second category includes representatives of things that served to
impress or amuse people—especially the wealthy. This was the case with
various animal automata in gold or silver that the mechanician Leon the
Philosopher made for Emperor Theophilus of Byzantium (ninth century)
to ornament "Solomon's throne" in the reception hall for foreign ambas-

sadors. There were also the horses and coachman that François-Joseph de Camus produced for the amusement of Louis XV as a child (seventeenth century). Finally, there was the false automaton but real chess player with which Wolfgang von Kempelen (eighteenth century) and then Johann Nepomuk Maelzel (nineteenth century) astonished Europe and America.

Machines in the third category concern particularly the field of bionics, but in order for their inventors or constructors to test hypotheses they formulated about the mechanisms of the living, various philosophical and religious prohibitions had to be lifted. Thus, at the end of the fifteenth century, Leonardo da Vinci dared to carry out human dissections, as did his distant Greek and Muslim predecessors, because the practice was not yet officially permitted in the West. No doubt thanks to such experiments, around 1495 he conceived (but probably did not construct) an automaton with the appearance of a German Italian knight who could sit down, stand up, move its arms and head. It could also open and close its jaw, possibly to emit sounds to accompany automatic percussions. This automaton presented two independent mechanical systems, one for the upper body and the other for the lower body. A mechanism based on notched pulleys situated in the chest could coordinate the movements of the shoulders, elbows, wrists, and hands. An external crank moved the legs, articulated at the hips, knees, and ankles, with the aid of cables representing ligaments. A reconstruction of this android automaton made from Leonardo's drawings may be seen at the San Diego Museum of Man.

However, it's only after René Descartes's mechanistic vision of animal behavior (seventeenth century), followed by Julien Offray de La Mettrie's view of human behavior (eighteenth century), that automata designers were really motivated to reproduce the living. Great technological advances notably linked to the development of watch making and its corresponding miniaturization were also necessary to help them realize this goal. The most celebrated craftsmen were Jacques Vaucanson and two men named Jaquet-Droz, father and son (eighteenth century).

Two surgeons, Claude-Nicolas Le Cat and François Quesnay, encouraged Vaucanson to set about a reproduction of means capable of reproducing the experimental intelligence of a biological mechanism. In 1738, Vaucanson presented at the Royal Academy of Sciences an essay describing the three automata that made him famous: the Flute-Player, the Galoubet and Tambourine-Player, and the Digesting Duck. The latter has remained the most famous of the three, described in the subtitle of the essay as "an artificial duck that eats, drinks, digests, and voids itself, cleans its wings and feathers, imitating in various ways a living duck."[2] A cylinder with

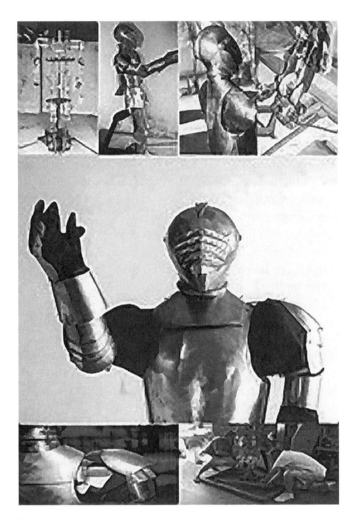

Figure 5.2
Reconstruction of Leonardo da Vinci's android at the Museum of Man in San Diego.
(© San Diego, Museum of Man.)

metallic cams worked levers that transmitted a movement by the intermediary of wires. With such cams, Vaucanson was able to program sequences
of gestures that made the fowl's behavior very realistic. The morphology
was very carefully rendered, each wing including more than four hundred
articulated pieces. The duck's physiology was the focus of virulent debates
at the time over the mechanical or chemical nature of digestion; as the
inventor put it, the duck digested "by a chemical dissolution" of the grains

it pecked. It then ejected the product of this transformation after its transit ended in an artificial anus.

The existence of this arrangement was much later contested by Jean-Eugène Robert-Houdin, as well known for his ability as an illusionist and mechanic as for the inaccuracy of his writings. Having had to repair the duck when it was presented at the Exposition des produits de l'industrie française in Paris in 1844, he claimed to have noticed that by a simple artifice it from time to time ejected a greenish substance prepared in advance, without any relation to what the duck ingested. This verdict was given more than a hundred years after the duck had been built, and, as Alfred Chapuis and Edouard Gélis remarked later, "we know how much Vaucanson's automata aroused in his day the emulation of copyists, and so we think that the duck described by the celebrated prestidigitator is one of these replicas. It seems to us that the trickery denounced by Robert-Houdin was much too crude and unworthy of the mechanical talents of the genius inventor, and finally does not conform to the description of the functions that he gave of them. We also know that the extreme

Figure 5.3
The automaton duck. Excerpt from the essay presented by Jacques Vaucanson at the Royal Academy of Sciences in Paris in 1738.

abundance of the ideas of the author of *Confidences of a Prestidigitator* led him to make assertions whose exactitude was impossible to check."[3]

Then, all trace of the duck and its secrets of fabrication were lost in the nineteenth century.

From Programmable Automata to Computers

Having produced these famous automata, Vaucanson was elected to the Royal Academy of Sciences of Paris thanks to support from King Louis XV. He then accepted the post of inspector of silk manufactures, which allowed him to use his cam cylinder to program designs for the weaving machines, opening the way, fifty years later, for Jacquard loom. In turn, the latter inspired Charles Babbage's calculator, which—according to his assistant Ada Lovelace—"wove algebraic models the way a Jacquard loom weaves flowers and leaves."[4] This machine is the ancestor of the computer.

Figure 5.4
The three famous automata of the Jaquet-Droz clockmakers. Left to right: The draughtsman, the lady musician, and the scribe. (Musée d'art et d'histoire, Département des arts appliqués, Neuchâtel, Switzerland.) See color plate 6.

Other eighteenth-century watchmakers, the Swiss Pierre Jaquet-Droz, his son Henri-Louis, and his adopted son Jean-Frédéric Leschot, were also associated with progress in programmable automata. The two to six thousand pieces that make up their three most famous automata—a girl playing music and two boys drawing and writing—are moved by even more sophisticated gears than Vaucanson's.[5] It is thought that these androids were realized in close collaboration with surgeons from the region and that their life-size skeletons were modeled on human skeletons. Their *moving anatomy* enables them to "really" play the organ or to "really" draw and write by tracing figures or letters with a goose quill, unlike more recent automata that only "seem as if" they are executing these activities. The height of realism, they all appear to breathe while performing, and a subtle detail noted by a doctor at the time is that "the musician breathes like women do, from top to bottom!"[6]

However, for lack of more such gifted inventors and mechanics, the fabrication of biologically realistic automata then fell into abeyance.

Protorobots

It was at the end of the nineteenth century and the start of the twentieth, when psychology started to distinguish itself from philosophy, that interest was reborn in testing living mechanisms by means of artificial systems. The innovative work of physiologists such as Ivan Pavlov and Jacques Loeb and of psychologists such as Edward Thorndike, John Watson, and Burrhus F. Skinner stressed the importance of perceptions, on the one hand, and of the association between perception and action, on the other. Several hypotheses concerning learning by "stimulus/response" were competing with each other, and some researchers thought of testing them on subjects that were not animals—machines. Psychologist Clark Hull called this approach "the robot approach"[7] when he advocated the use of machines in psychology as "a form of prophylaxis against subjectivist anthropomorphism."[8] It was in this context that the first robots made their appearance.

Unlike automata, which always reproduce the same behavior once they are switched on, robots can act differently as a function of the context in which they find themselves. They are in effect endowed with sensors, the equivalents of sensory receptors that apprehend environmental stimuli, both external and internal.[9] A *control architecture*, the equivalent of the nervous system, treats these stimuli and decides as a consequence what actions the actuators (like muscles, for example) will perform.

In 1912, engineers John Hammond Jr. and Benjamin Miessner constructed "Electric Dog"—an electromechanical dog devised as an "artificial heliotropic machine" with which it was possible to test the physiologist Loeb's theory of tropisms. The dog was a box furnished with two sensors in front and with three wheels (two motorized wheels in front and a swiveling wheel behind) that moved it by "phototropism": it was capable of directing itself toward the light of a flashlight even if the light changed direction. Its control mechanism was very simple: the front wheels turned faster when the sensors were more stimulated; the rear wheel had an electromagnet that oriented it so that the box moved toward the most stimulated sensor. The effect was spectacular: it was the first time that a machine moved without precisely predetermined motor programs. A magazine even described this machine as "to inherit almost superhuman intelligence,"[10] and Loeb, very impressed, generalized his theory of tropisms to nonmotor faculties, such as curiosity and pleasure.

In this heroic period, let us also mention the engineer Bent Russell (1913), the psychologist John Mortimer Stephens (1929), and the electrochemist Thomas Ross (1935), who aimed to increase protorobots' faculties by means of control architecture able to *learn*. Hypotheses about "trial-and-

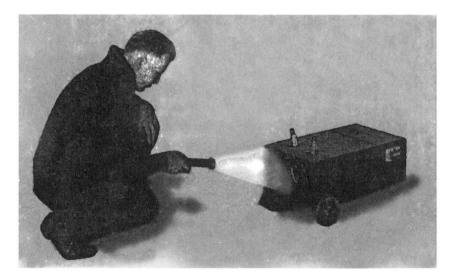

Figure 5.5
John Hammond Jr. and Benjamin Miessner's "electromechanical dog" drawn to light from a flashlight. (In B. F. Miessner, *Radiodynamics: The Wireless Control of Torpedoes and Other Mechanisms* [New York: van Nostrand, 1916], 197.)

error learning" proposed by behaviorists such as Thorndike and Watson could be tested, thanks to these machines, according to the idea that electric transmission might represent conduction in the nervous system and that a reduction or increase of resistance might correspond to the effects that rewards or punishments, following good or bad responses, have on the organism.

Many protorobots of this type followed, as quickly forgotten as the preceding ones because of the limits of both the theoretical knowledge and the technical expertise on which they were capitalizing.[11] They left behind the idea that living things' complex behaviors can be reproduced only by mechanisms that are themselves extremely complex. But this notion was to some extent disproved by the machines that followed.

Cybernetic Robots

The Second World War precipitated research into so-called intelligent machines. A parallel was established between mechanisms that allowed both living systems and artificial systems to pursue predefined goals, and great importance was now granted to the *internal state* of machines. These machines were no longer reduced to simple stimulus-response systems; while anticipating the desired states, they could be capable of controlling themselves.

This idea is at the heart of cybernetics (from the Greek *kubernêtikê*, "the art of steering a ship"), a discipline that Wiener developed after 1942, with many representatives of other disciplines (information sciences, psychology, sociology, physiology, anthropology) following him. The mechanisms of positive and negative feedback were considered the basis of the autonomy of machines.[12] In 1943, the guidance system of navy missiles that were able to anticipate the trajectories of enemy planes to bring them down exploited these mechanisms. In the same year, neuropsychiatrist Warren McCulloch and mathematician Walter Pitts published an article describing artificial neurons able to implement the laws of formal logic. These works oriented cybernetics on a path even more closely linked to biology than it had been before, strengthening the metaphor of a control architecture that was analogous to the human brain.

New machines were constructed to apply the principles of autonomous control. The most representative were the Homeostat ("ultrastable machine") by the psychiatrist William Ross Ashby (1948), the cybernetic turtles of the neurophysiologist William Grey Walter (1949), and the electronic fox of Albert Ducrocq (1950–1953).

Cybernetics

In the 1940s, MIT mathematician Norbert Wiener and engineer Julian Bigelow founded a discipline that studies the similarities and differences between biological processes (which are monitored by the nervous system) and technical processes (which are monitored by mechanical, electrical, or electronic means) when each of them is tending toward a previously defined state or goal. The common principle of functioning rests on the notion of *homeostasis* (remaining the same), which explains how a system produces behavior capable of reducing the difference between its current state and the state in which it will be when it has reached its goal. Various mechanisms of retroaction of effects on causes (feedback) are relied upon to put this principle to work.

The two researchers got the idea while observing the behavior of missiles that are based on these principles. These systems seemed to be "intelligent" because they proved to be capable of anticipating the trajectory of their targets, but they seemed to suffer from motor troubles because when one tried to reduce air friction along their trajectory, they entered a series of uncontrollable oscillations. Physiologist Arturo Rosenblueth coincidentally noticed a similar pathology among patients suffering from a lesion of the cerebellum: they were in effect able to program the overall trajectory of a glass of water to their mouths but could not execute it correctly, for their movements were amplified in such a way that water spilled from the glass before the patient completed its trajectory. Wiener then had the idea of adding, on top of the principal retroaction loop that interacts with the external world, a secondary loop interacting with the internal milieu and evaluating at every moment the correct execution of an action.

This general principle has been extended not only to the behavior of individual and machines, but also to society and industry. In a book published in 1950, *Cybernetics and Society*, Wiener even warned political leaders of the impact of this science, foreseeing that the replacement of workers by intelligent machines might herald the end of democracy if leaders neglected to prepare the postindustrial evolution that would allow people to live entirely liberated from working.

Figure 5.6
Cybernetic robots. Top left: William Ross Ashby and the Homeostat, the "ultrasta-
ble" machine. (In Pierre de Latil, *Introduction à la Cybernétique. La Pensée Artificielle*
[Paris: Gallimard, 1953], 225). Top right: Grey Walter and ELSIE, his cybernetic
turtle, going back into its "nest." (© The Burden Neurological Institute, Bristol, UK.)
Bottom: Job, Albert Ducrocq's electronic fox, is currently stored in the reserve of the
Musée du Conservatoire national des arts et métiers in Paris. (© Musée du Conservatoire National des Arts et Métiers/M. Favareille.)

The first is an apparatus composed of four galvanometers with interconnected moving electromagnets. The needle of each of these galvanometers is supposed to remain between two boundaries, and the "survival" of the Homeostat depends on whether it manages to maintain each of its needles within the associated *viability zone*. The internal organization of the apparatus is such that when the experimenter moves one of the needles to pull it beyond the assigned limits, all the needles start to move because the whole system resists external perturbation and reorganizes itself to bringing all the needles back to their respective zones of stability. This reorganization puts two feedback loops to work: one is charged with rectifying the disequilibrium of the needle that was initially perturbed, and the other is charged with systematically seeking a position of equilibrium suitable for all the needles. Although from the standpoint of its behavior this machine was not as attractive as John Hammond Jr. and Benjamin Miessner's phototropic dog, it opened new perspectives for understanding the principles of living systems' adaptations. The means enabling an animal's survival in a multiconstraining environment could henceforth be interpreted in terms of a constant search for compatibility between homeostases that balance external and internal states.

William Grey Walter took up this perspective but applied it to mobile robots. This gave him the opportunity to make friendly fun of Ashby's immobile box, calling it the *Machina sopora*—the sleeping machine! His two robots, each covered with a carapace, belonged to the species *Machina speculatrix*—or machine that "seeks a goal"—and were nicknamed ELMER (ELectro MEchanical Robot) and ELSIE (Electro, Light-Sensitive Stabilized Internally and Externally), with ELMER being ELSIE's prototype. These cybernetic turtles have a dual historical interest. On the one hand, they are the first robots whose behavior was not totally predictable by humans because they seemed to pursue their own goals. On the other hand, they used the first vacuum tubes, semiconductors, and transistors—in other words, the first components of electronics, which are much faster than electromechanical components.

Both turtles possessed a light sensor, a shock sensor, two wheels in back, and a directional wheel in front. A basic circuit helped maintain perception of median luminosity: when the luminosity was too strong, the machines slowed down; when it was too weak, they "sought" (by exploration) the proximal light source. This effect was combined with a monitoring of their internal state, so that light was aversive when their batteries were full and more attractive when they were empty. Thus, the turtles naturally went to their strongly illuminated nest when it was time for them to recharge. Thanks to their capacities, the two machines could evolve for a long time without being depleted of energy. Moreover, when they were fitted with a headlight fixed to their carapaces, they seemed to be attracted to each other; they deployed behavior of successively approaching and stopping, which Walter characterized as "courtship behavior" similar to those of two pigeons. The apparent complexity of their behavior was all the more astonishing because it called upon simple homeostatic principles.

Just when protorobots had been abandoned because the mechanisms underlying the behaviors of the living were considered too complex, these turtles brought proof that complex behaviors might emerge from simple mechanisms.

Walter conceived of other robots, including the COnditioned Reflex Analogue (CORA)—or *Machina docilis*, a machine that learned easily by trial and error. CORA was given a form of memory and was capable of learning an association between a neutral stimulus and a response, like Pavlov's dog, by operant conditioning. Nevertheless, this machine was not mobile like the two preceding ones. It still exists and can be found at the Science Museum in London. Meanwhile, researcher Owen Holland has reconstructed ELSIE in Bristol.

French scientist Albert Ducrocq created Job, the electronic fox, shortly thereafter and presented it in 1954 at the French exhibition Life in the Year 2000.[13] Unlike its predecessors, this robot was equipped with several sensory modes: a visual system (two photoelectric cells), an auditory system (a microphone connected to an amplifier), two tactile systems (shock and obstacle avoidance sensors), and a vestibular system (sensors placed in the "neck" that gave a sense of orientation). Made of cardboard, wood, electronics, and covered with fox fur, it possessed capacities of learning similar to CORA's, thanks to a memory of magnetic bands, and could even communicate by means of two lamps, one red and one green, situated at the top of its head. Given this robot's capabilities, its inventor called it a "cunning apparatus, for thanks to five sensory channels it was possible for it to obtain a rough model of its environment, constantly reviewed and corrected by the lessons of experience."[14]

Why weren't more such robots made and perfected in the following years? No doubt because of a coming wave known as "artificial intelligence" and its almost exclusive interest in the human brain.

Intelligent Systems

In 1936, the British mathematician Alan Turing conceived the scheme for a machine able to handle information; he demonstrated that it could theoretically perform the same operations as the human brain—or as any other apparatus that could also handle information.

In parallel to the cybernetic metaphor, a competitor known as the "computational metaphor" began to take off: it was no longer the living nervous system that would inspire the control architecture of an artificial system, but instead the computer that would impose a new way of speculating about and interpreting the human brain. In effect, the computer had been sufficiently refined, although the equivalent of one of today's laptops occupied the space of several rooms! It was considered to be capable—like a human being—of dealing with data coded in the form of symbols (at the time "0" and "1") by two memories—random access memory (RAM) and read-only memory (ROM), corresponding respectively to short-term and long-term memories.

In 1956, in the course of a summer conference at Dartmouth College in New Hampshire, the pioneers of artificial intelligence—including Marvin Minsky, John McCarthy, and Claude Shannon—were engaged in finding "how to make machines use language, form abstractions and concepts, solve kinds of problems now reserved for humans, and improve themselves."[15] In particular, researchers implicitly admitted that artificial systems might do without sensory and motor organs in order to develop cognitive capacities that were as elaborate as those of humans. Spectacular applications quickly seemed to support their point of view. Programs such as the General Problem Solver[16] were capable of resolving many logical problems, ranging from simple conundrums such as the Hanoi towers[17] to the demonstration of complex mathematical theorems. Systems such as Dendral and Mycin were developed,[18] which proved expert in the identification of chemical molecules and diagnosing of blood diseases, respectively. They could even furnish explanations for their decisions, which human experts in these fields were sometimes incapable of doing. Joseph Weizman's Eliza program (1965) used the therapeutic method of the American psychologist Carl Rogers to have a conversation in natural language with a human interlocutor—a performance that lost its mystery once one understood that the program responded only with phrases made up of key words used by patients or that Eliza simply repeated a sentence when she had no answer to give.

In 1972, though, at the height of these so-called intelligent systems, the American philosopher Hubert Dreyfus, in his book *What Computers Can't Do: A Critique of Artificial Reason*,[19] denounced the serious limitations of these disembodied brains—a defect he attributed essentially to the fact that programmers must furnish to artificial systems a priori a very great amount of human knowledge for these systems to be able to function effectively. In particular, the meaning attributed to symbols manipulated by the computer is imposed by the programmer. During the constitution of this "intellectual baggage," humans will naturally neglect to give machines knowledge that appears so obvious—such as the knowledge that a heavy object does not stay in the air by itself or the knowledge that if the president is in the White House, his right foot is there too. But such implicit knowledge is crucial to resolve the concrete problems of everyday life or even to understand a written text or a speech. Also called *common sense*, this knowledge is not available at birth but must be elaborated under the effect of experience by an organism equipped not only with a brain, but also with a body endowed with sensory and motor organs.

It proved impossible for a disembodied system to acquire common sense. In 1984, two young MIT researchers, Doug Lenat and Ramanathan Guha, wanted to accept the challenge. Their program, Cyc, aimed to be able to converse in natural language with humans and to master all the subtleties of the language. Cyc had to *infer* its common sense on the basis of knowledge of all kinds that was furnished either by interaction with numerous people of all ages, sexes, conditions, jobs, and social situations or else by analysis of the content of various encyclopedias. Cyc became able to understand the meaning of the very ambiguous phrase "Can a can cancan?" but it appeared that the quantity of information necessary for the correct understanding of a conversation undergoes an exponential growth, and that the assessment of Cyc's "human intelligence" had to be revised downward.[20] The ambitions of this program were appreciably reduced and reoriented to the extraction of data on the Internet, a task for which it proved very competent. At the start of the 1990s, most of the programs previously characterized as "intelligent" and were similarly reconverted. "Strong" artificial intelligence is now called "good old-fashioned artificial intelligence" (GOFAI).

Programs of this type are of course still developed today, but they aim to deal "intelligently" with all kinds of data, and they no longer claim to use the same mechanisms as a human mind. For example, Deep Blue and Deep Fritz, which have beaten in official competition two world chess champions, Garry Kasparov and Vladimir Kramnik, are able to calculate millions of chess moves per second and to memorize the unfolding of thousands of matches, but they reason in no way like these champions.

It is also because of the lack of "common sense" that the few robots constructed in the heroic GOFAI era were scarcely capable of evolving in the real world. Their conceivers had to provide them with so much

Disembodied Intelligences: Eliza and Cyc

A conversation between a young woman (YW) and Eliza, the "cybertherapist":

YW: My boyfriend says I'm depressed much of the time.

Eliza: I am sorry to hear you are depressed.

YW: It's true. I'm unhappy.

Eliza: Do you think coming here will help you not to be unhappy?

YW: I need some help, that much seems certain.

Eliza: What would it mean to you if you got some help?

YW: Perhaps I could learn to get along with my mother.

Eliza: Tell me more about your family.

YW: My mother takes care of me.

Eliza: Who else in your family takes care of you?

YW: My father.

Eliza: Your father.

YW: You are like my father in some ways.

Eliza: What resemblance do you see?

YW: You are not very aggressive, but I think you don't want me to notice that.

Eliza: What makes you think I am not very aggressive?[a]

It is said that the vice president (VP) of Bolt, Beranek, and Newman, who would later be one of the developers of the Internet, thinking he was conversing with a colleague by teletype, was in fact connected to Eliza. The following dialog ensued:

VP: Do you think that I can use the computer this morning?

Eliza: Why do you ask?

VP: I might be able to make some additional Telcom sales.

Eliza: Why aren't you so sure?

VP: My perspective customers have never seen a demonstration of the system.

Eliza: Does this mean very much to you?

VP: Of course it does.

Eliza: You seem very positive about that.

VP: This is ridiculous.

Eliza: What do you mean, this is ridiculous?

VP: Is there any reason why I can't use the computer now?

Eliza: Tell me more about your prospective customers.

VP: Please dial me up on 491-1850.

Of course, the computer didn't call him. This so infuriated the vice president that he called his "colleague" again and said: "Why are you being so snotty to me?" The imperturbable colleague answered: "What do you mean, I am being snotty to you?"[b]

Cyc is programmed to reason by inference; some of its deductions from information given by humans include:

- Mothers are always older than their children.
- To eat one has to be awake.
- One can see people's noses but not their hearts.
- When people die, they remain dead.
- If one cuts a piece of butter in two, one obtains two pieces of butter, but if one cuts a table in two, one does not obtain two tables.
- When Fred holds his razor, he is composed of flesh, metal, and electrical components, but he is not a machine.
- When President Lincoln was in Washington, his left foot was too; his name and birth date have not changed in the course of his life.

Note

a. In Joseph Weizenbaum, *Computer Power and Human Reason* (San Francisco: Walter Freeman, 1976), 369.

b. The anecdote comes from Daniel Bobrow, who wrote STUDENT, one of the earliest natural-language processing programs, for high-school-level algebra problems.

information about the physical world encoded by appropriate symbols and then decode the productions of their internal reasoning that it seemed unrealistic to count on these machines' intelligence alone to perform basic tasks concretely.

Adaptive Robots—the Animats

The cause of the relative stagnation of intelligent systems can be summed up as follows: what is difficult for a human to do is quite easy for a computer and vice versa! For a computer, to rediscover Thales's theorem or to calculate the best moves in chess is easier than to learn to locate a useful object or to avoid the obstacles in a room. In other words, to reason intelligently about the real world is much easier than to interact with it.

In 1986, a young Ph.D. from John McCarthy's laboratory, Rodney Brooks, braved the dominant ideology within this temple of artificial

Figure 5.7
The robot Genghis, pioneer animat. (© Rodney Brooks, MIT Computer Science and
Artificial Intelligence Lab.)

intelligence and decided to conceive of robots not as intelligent, but simply
as *adaptive*—that is to say, potentially able to get by in the real world. To
do this, he aimed first to reproduce the capacities that we humans share
with animals, not those that are specific to us. Like them, we know how
to move, to orient ourselves in an environment, to fulfill our essential
needs—all the basic activities without which we would be incapable of
developing the least abstract intelligence. Brooks reintegrated action—the
great absence in strong artificial intelligence—into the conception of his
robots, which were more inspired by insects than by humans. He did not
supply them with symbols whose meaning was predetermined, but instead
equipped them with sensors to deal with brute data from the environment.
He did not program complex reasoning processes, but rather simple
modules of behavior that were activated in parallel in close interaction
with the environment.

It is upon these simple principles that his hexapod robot Genghis
became capable of avoiding obstacles without having the least representation of the concept of "obstacle" and of leaving a room through the door
without having the least representation of the concept of "door"—notions
that, in contrast, were absolutely indispensable for its intelligent
predecessors.

A community of researchers is being recognized for this approach,
which is called the *animat approach* (the term *animat* is a contraction of
artificial and *animal*).[21] Animats are autonomous, adaptive, and situated.

Autonomous, because they have needs and motivations to satisfy in order to survive and they do not receive, at least in principle, any human aid to do so. *Adaptive*, because they also have capacities to learn and to evolve, which contribute to increasing their chances of surviving and accomplishing their mission in a more or less changing, unpredictable, or threatening environment. *Situated*, because their faculties of adaptation are rooted in the sensory-motor loops induced by their permanent interactions with their milieu.

The animat approach draws heavily on the knowledge that biologists have acquired of living systems. In return, it aims to confirm or to improve that knowledge, if not to generate new hypotheses. A half-century later, this approach mixes the spirits of those who once conceived of both protorobots and cybernetic robots (as discussed earlier), but it also capitalizes on information and electronic technologies more advanced than the previous electromechanical circuits.

Sources of Inspiration Aplenty

Since the appearance of life on the planet about 3.5 billion years ago, there has been proof that there is no single way of surviving. Apart from the *extremophiles*[22]—those microorganisms that can resist temperatures of −15 or +110 degrees Centigrade, radiation 1,500 higher than what humans can bear, and gravity of 1 million g's, equivalent to a depth of 11,000 meters under the ocean—most organisms have found multiple ways to occupy ecological niches under the more normal physical conditions that are our own. Organisms are very different from each other with respect to their anatomies, their motor and sensory organs, and their control systems, but all have proved capable of reproducing generation after generation and of occupying milieus of the utmost diversity.

Because these characteristics have demonstrated the aptitude to guarantee the survival of the organisms in question, they have inspired engineers who are looking to conceive of adaptive animats. Some of the results obtained in this way are described in subsequent chapters.

6 Effective Actuators

Nature acts, man makes.
—Immanuel Kant (1920)

The invention of the wheel is often advanced as an argument that nature has not invented the most effective method for living organisms to move. However, two remarks suggest that this arrangement was not the most adaptive: on the one hand, the first organisms moved in water, and, on the other, when animals colonized terra firma about 400 million years ago, paths were not as well prepared as they were 5,500 years ago, when the Sumerians appear to have invented the wheel.

The best means of moving in a liquid element or over a chaotic and muddy surface encumbered with branches and stones are swimming, crawling, and walking.

Swimming

Thunniform

The different ways of swimming have been abundantly studied, and the principal types of propulsion have been identified.

In *anguilliform* swimming, like that of the eel or moray, the whole body is implicated in the undulation movement. In *carangiform* and *thunniform* swimming, like the way trout or tuna do, the rear part of the body oscillates and plays a role in propulsion, the pectoral fins serving as rudder. It is the reverse in *ostraciiform* swimming, like the boxfish we mentioned in a preceding chapter does, for the whole body is rigid. The pectoral fins propel the animal by oscillating, while the caudal fin serves as rudder. The first three ways of swimming are costly in energy but allow fast sprints and rapid turns. The fourth is not costly in energy, but much slower. The morphology of the animals also contributes to the efficiency of this kind of movement.

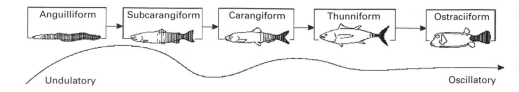

Figure 6.1
Different forms of swimming classified according to the proportion of the body (in gray) that undulates or oscillates during swimming. (© Dunod.)

Many swimming robots adopt these various modes of propulsion. The thunniform mode is the most sought after, for it would enable submarines to reach a speed and a maneuverability that they cannot currently attain. The red tuna, for example, reaches speeds of 75 kilometers an hour and can make instantaneous turns of 90 degrees.

Several micro- and macro-robots ranging in length from 50 millimeters to 3 meters were conceived after this model at the National Institute of Maritime Research in Tokyo. The prototype UPF-2001, the fastest, can reach 0.97 meters per second. Like the Korean PoTuna or the American RoboTuna of MIT or the Vorticity Control Unmanned Undersea Vehicle (VCUUV) at Draper Laboratory—which reaches a top speed of 1.2 meters per second and turns of 75 degrees per second— these robots are far from attaining the performances of their natural models, but they nevertheless allow us to understand how this mode of swimming can perform at high levels. The turbulence generated in the fish's wake is an important source of energy waste. The geometry of its body, the flexibility of its tail, as well as the double rapid beats of its caudal fin all contribute to minimizing this loss. Its pectoral fins, compensating for the danger of pitching, are also essential to maintain body equilibrium.

Plesiosaurus
The robot Madeleine, from a New York team directed by John H. Long, answers a more fundamental objective. Rather than looking for the fastest engine, its designers used it to test hypotheses about the modes of movement of aquatic tetrapods, animals with four flippers. Some species have disappeared, such as the plesiosaurus (literally "near the lizard"), but others are still living, such as penguins, sea turtles, and seals. Madeleine demonstrated that only the simultaneous use of all four flippers is capable of

producing significant acceleration, but that it entails a prohibitive energy cost. It is for this reason that the plesiosauruses, known for being the most formidable predators of the Mesozoic era, have disappeared: using their four flippers to throw themselves on their prey, they could not manage to compensate for the consequent energy depletion.

Madeleine has also demonstrated that at a normal speed, the use of the two front flippers alone is as effective and much more economic than the application of all four at the same time. It is also a more effective way to avoid obstacles. One may conclude that evolution seems to have retained this mode of propulsion, for it is the one used by all the aquatic tetrapods still alive, where the hind members serve only for steering.

Aquarium Fish

Will the increasingly faithful reproduction of the behavior of fish permit us to envisage the gradual replacement of animals in aquariums? Some people are actively working on this.

The Japanese firm Mitsubishi Heavy Industries has constructed a robotic coelacanth that is 120 centimeters long and able to swim at 0.3 knots—which can be rented for the modest sum of 1,200,000 yen a month, or about $10,000. Doing so would prevent the poaching of these rare fish, which were believed extinct for millions of years, but of which there are still a few living examples in Indonesia.

The London aquarium has recently exhibited three robot carps, designed by researchers in the University of Essex computer department under Jindong Liu and Huosheng Hu. According to them, G9-1, G9-2, and G9-3 aim to acquire "the speed of the tuna, the acceleration of the pike, and the ease of the eel." For the time being, each 50-centimeter-long carplike body swims in an autonomous way—that is, without being remotely controlled—for 5 hours and at a rate of 30 centimeters per second. They are equipped with sensors that enable them to avoid glass walls and other fish. The current research aims to improve these robots' autonomy so that they can recharge themselves at electrical sources, as if they were going to get food. They might also communicate with each other—silently, like the natural carp.

Their behavior is so realistic that visitors have trouble imagining that they are not real, although their appearance does not really lend itself to this confusion! By enticing whole schools to visit the aquarium, researchers hope that these artificial creatures will give the youngest students a taste for scientific careers.

Figure 6.2
One of the robot fish in the London Aquarium. (© Dr. Jindong Liu, University of Essex.) See color plate 8.

Crawling

Lateral, Accordion Style, Linear

Anguilliform undulations and terrestrial crawling do have points in common: all the vertebral segments of which the body is composed collaborate with each other in effective locomotion. Robotic research in this realm consists, therefore, of understanding how this coordination is organized.

Many robot snakes at Carnegie Mellon University aim to reproduce the main ways of crawling: *lateral* to progress over flat terrain, *accordion style* to climb, and *linear* to move over irregular terrain. Various robots deploy all sorts of functions by various preprogrammed cyclical movements, for apart from classic crawling over various surfaces, they can swim, climb along walls, climb up a column by circling it, and edge their way into narrow conduits. Researchers even envisage miniaturizing these engines:

Figure 6.3
Omnitread, the all-terrain snake robot. (© Johan Borenstein,University of Michigan.)

by gliding them into blood vessels or organic tissues, they might examine wounded people before they are transported to hospital, or they might use them for noninvasive surgery.

Omnitread, the robot developed at the University of Michigan, serves more fundamentally to understand how to optimize crawling movements. Conceived as a vertebral column controlled by a sort of spinal cord, it does not have preprogrammed movements like the previously described robots. A single motor placed in the central segment has proved more effective than several motors divided among several modules, enabling it, solely by interaction with the environment, to perform a crawl that automatically adapts to any type of terrain.

Walking

Locomotion with legs is faster and better adapted to overcoming obstacles than crawling, a fact that various robots—octopod, hexapod, quadruped, and biped—are demonstrating in various laboratories.

Hexapod

Hexapod walking is the most stable. The stick insect manages to adapt to various rough terrains, even when it has unexpectedly lost one of its six legs. A professor in the biology faculty at the University of Bielefeld, Holk Cruse, after having studied the principle of motion control in this animal, constructed a robot that demonstrates that there is no centralized oscillator to command the periodic phases of its walking. On the contrary, each leg is autonomous and interacts only very locally with its neighbors. Thus, when one of them is amputated, local interaction is reorganized among the nearest legs. The stick insect's cycle of movements—swinging forward, settling on the ground, pushing back, swinging forward—emerges reflexively though interaction of the neuromuscular system with the ground.

Similarly, in a laboratory at Case Western University, it has been shown that speed of walking can be accentuated by giving robots, like cockroaches, legs with different possibilities of nonconstrained movement in space—called "degrees of freedom": five for the forward legs, four for the middle legs, and three for the rear legs.

In the Performance Energetics and Dynamics of Animal Locomotion (PolyPEDAL) laboratory at Berkeley, researchers have also used the motion principles of these insects to conceive the Sprawl series, the most rapid of hexapod robots. They are still ten times less swift than their models, the cockroaches, which move at fifty times the length of their body per second!

Quadruped

Among the quadruped robots, let us mention the Stickybot, perfected by Mark Cutkosky and Kim Sangbea at Stanford, in collaboration with those working on the gecko mentioned in an earlier chapter. This robot's digits are covered with a hairy nanostructure that enables it to climb along a pane of glass or up a smooth wall, thanks to the van der Waals force, the famous dry adhesive compatible with any support. The originality of this covering resides in the fact that the nanoasperities possess a flat and oriented face, a little like the spatulates of the gecko. This enables the digits to attach and detach themselves much more rapidly than with the procedures elaborated previously.

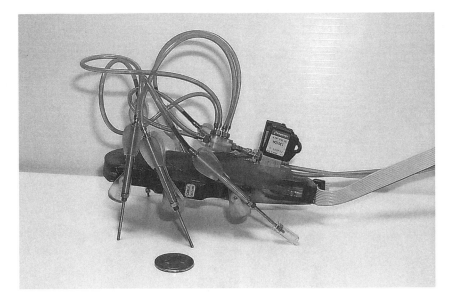

Figure 6.4
Sprawlita, one of the fastest artificial creatures, inspired by the cockroach. (© Mark R. Cutkosky, Stanford University Center for Design Research.)

Let us also mention BigDog, which Boston engineers conceived for the U.S. Army and did not hesitate to call the "most advanced quadruped robot in the world." This robot can reach a speed of 5 kilometers per hour on flat ground, go up an inclined plane, and carry a load of 50 kilograms.

Biped
In order to make a biped robot walk in the same way as a human—which constitutes one of the great current challenges of robotics—it would have to cope with phases of disequilibrium, and to do that it would have to be able to perceive the position of its body, thanks to proprioceptive sensors, as well as the state of its environment, thanks to exteroceptive sensors (visual or tactile).

Little progress had been made in this domain before a French-American team proposed in 2002 a mathematical method allowing the choice of a strategy of command most adapted to the robot's morphology. The biped robot Rabbit, from Grenoble's Automatics Laboratory, its trunk provided with two legs that have no feet, took advantage of this command strategy. Other successful walking robots have been produced since then, such as

Figure 6.5
The Gecko robot and its nanosetulae. Left: Stickybot, the quadruped robot gecko able to climb a pane of glass, watched by its creators. Right: The pulvilli are covered with setulae ending in spatulates. (© Mark R. Cutkosky, Stanford University, Center for Design Research.) See color plate 9.

Figure 6.6
ASIMO, the humanoid biped able to run and climb stairs. (© Honda Motor Europe Ltd.)

Plate 1

Plate 2

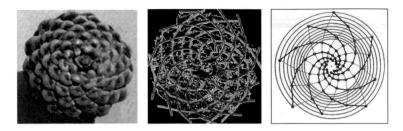

Plate 3

Plate 4

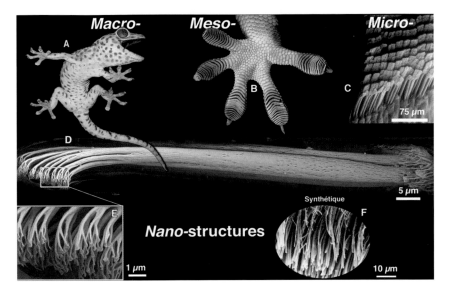

Plate 5

Plate 6

(−60,0)   (−90,0)   (80,0)

(−45,0)   (0,0)   (45,0)

(−30,0)   (0,−60)   (30,0)

Plate 7

Plate 8

Plate 9

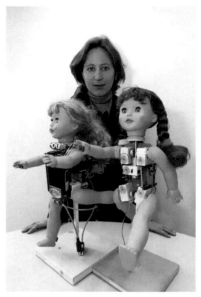

Plate 10

Plate 11

Plate 12

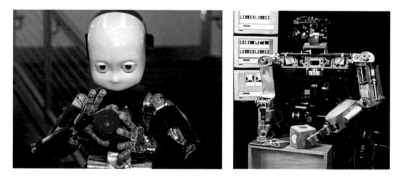

Plate 13

Plate 14

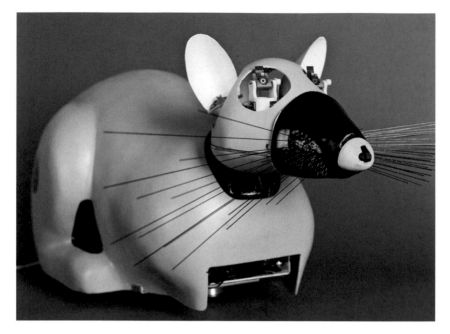

Plate 15

Plate 16

Plate 17

Plate 18

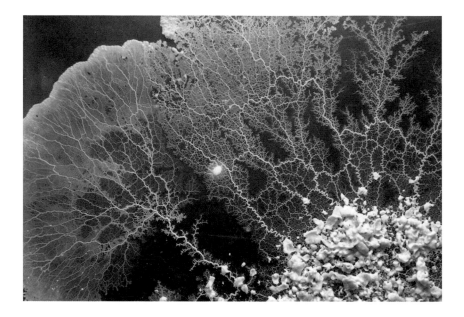

Plate 19

Plate 19 (continued)

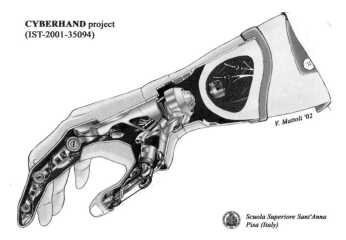

CYBERHAND project
(IST-2001-35094)

V. Mattoli '02

 Scuola Superiore Sant'Anna
Pisa (Italy)

Plate 20

Plate 21

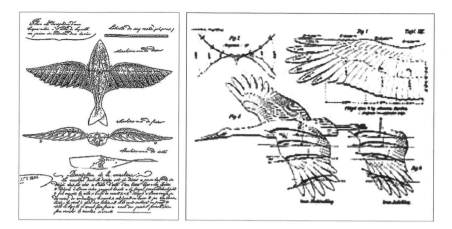

Figure 6.7
Left: The albatross of Jean-Marie Le Bris. Right: Model of the stork by Otto Lilienthal.
(© Dunod.)

the American RunBot, which moves at three and a half times the length
of its legs per second, or the Japanese Advanced Step in Innovative MObility (ASIMO—pronounced *ashimo*, meaning "legs also" in Japanese), which
reaches a speed of 5 kilometers per hour, climbs stairs, and can do a few
dance steps. In its latest version, it is even able to run.

Flight

Flight is a technological challenge that is otherwise complex for an artificial
system, for it must evidently act constantly against gravity.

Many nineteenth-century airplanes were inspired by the forms of birds
or mammals, such as the albatross of the navigator Jean-Marie Le Bris,
the stork of the Lilienthal brothers, and the bat of Clement Ader. Although
totally supplanted for a very long time by airplanes that are not biomimetic and have fixed wings, engines with beating wings are today enjoying an abrupt resurgence of interest, notably in the drone industry. Such
flying engines without pilots would offer various advantages, such as the
ability to fly at low speed, to veer abruptly, and to practice stationary
flight.

Insects and birds' various flying techniques are thus the object of much
attention. By beating their wings, insects are borne aloft half by pushes
upward and half by pushes downward. Birds achieve this movement only
through wing pushes downward—except for the kolibri hummingbird,

which, with a size similar to a large insect, uses an intermediary technique composed of three-quarters of a push downward and one-quarter of a push upward.

The Kolibri

This bird was the inspiration for Mentor, the ornithopter with four remote-controlled wings developed at the University of Toronto by a team directed by James DeLaurier, although with its size of 30 centimeters and its weight of 500 grams, it is much larger than this animal. Nevertheless, it is the first artificial apparatus to have maintained a stationary flight for ten minutes by reproducing the hummingbird's "clap-fling" behavior. Yet its great consumption of energy considerably limits the duration of its performance.

Insects

Entomopter, being perfected at Georgia Tech's research institute by Robert Michelson's team, takes on the liquid fuel necessary to agitate its wings in the manner of a moth, at a frequency of 10 hertz. It still needs complex calculations to evaluate its aerodynamics and stability as a function of many parameters such as its weight, its wing structure, and the chemistry of its artificial muscles—the nature of the materials that compose a flying engine are in effect just as important as its morphology. This engine, with a wingspan of about one meter, is supposed to fly above the surface of Mars. The weak atmospheric pressure that surrounds this planet would oblige airplanes with fixed wings to fly constantly at 400 kilometers per hour, which would prevent their taking off and landing. Only light engines with beating wings can both fly to take photos and land to take samples.

Many designs for flying micro air vehicles (MAV) that imitate the flight of flies, bees, or dragonflies are in the works in several places in the world, most designed for military applications. The smallest engines are 10 to 25 millimeters long and can take cameras and microphones to perform reconnaissance missions.

Other bionic applications of flight will be described in a subsequent chapter.

Grabbing

To perform certain tasks, it is often indispensable to be able to grip or hold objects. Many artificial actuators are inspired very naturally by the

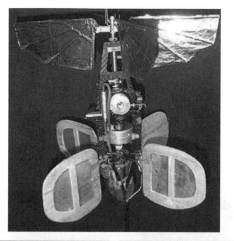

Figure 6.8
Bird and insect drones. Top: Mentor, the artificial kolibri. (© James DeLaurier, University of Toronto. Bottom left: Robert Michelson testing an arrangement of the beating wings of Entomopter, the artificial moth. (© Robert Michelson, Georgia Institute of Technology.) Bottom right: Extract from a video animation describing the use of Entomopter in the Martian atmosphere. (© NASA Institute for Advanced Concepts.) See color plate 17.

Figure 6.9
OCTarm, the "elephant" robot. (© Ian D. Walker, Clemson University.)

human hand, but other less-sophisticated forms have been envisaged. This is the case of robots in the sOft robotiC manipulaTORs (OCTOR) project, conceived by researchers at the University of South Carolina and modeled on octopus tentacles or the elephant's trunk. The OCTarm, for example, has an arm moved by compressed air that can pick up any unknown object thanks to pressure and position sensors. Its dexterity is superior to that of commercial robotic arms, but less amazing than that of its animal models. An elephant can catch a peanut as well as a tree trunk, and an octopus can delicately explore the recesses of rocks or unscrew a cork.

Perforating

The search for life on distant planets will most likely entail drilling, for ultraviolet rays have almost surely suppressed any life on the surface. From this perspective, British researchers are refining a perforating apparatus inspired by the wasp's ovipositor, which serves to dig into the bark of trees so that the wasp can lay the eggs there. The sort of needle of such an apparatus is made of two parts functioning side by side alternately, one that is equipped with teeth that perforate the substratum under compression so as to avoid buckling and one that is made of pockets that bring back to the surface the corresponding detritus. The biomimetic artificial system weighs less than a kilogram, consumes barely three watts, and can reach depths of one to two meters.

7 Amazing Sensors

No conjurer can equal nature: she performs before our eyes, in full light, and yet there is no way of penetrating her tricks.
—Rémy de Gourmont (1924)

Seeing

The Fly's Movement Detectors

For flying robots (like those mentioned in the preceding chapter) to be able to navigate successfully someday—including localizing, orienting, and steering themselves—they have to be given the ability to evaluate the distance that separates them from the ground and from any other obstacle. To do this, tools invented by engineers—such as radar or laser telemetry—may well prove unsuitable for such engines and for the applications for which they were designed.

A team of Marseilles researchers directed by Nicolas Franceschini has developed a navigation apparatus of a different nature inspired by the way the fly uses movement-detecting neurons located in their ommatidiae—the units that form its compound eye—to evaluate the speed of displacement on their retina (made up of photoreceptors at the back) of objects present in the environment. Such speed is called the *optic flow*. These neurons are particularly sensitive to the movement of objects because they can detect movements ten times more quickly than the human eye.[1] Because the amplitude of the optic flow depends on the distance of the perceived object, the higher the insect is, the more slowly the earth seems to move under it. Researchers presume that the fly adjusts its landing behavior according to the current value of the optic flow, seeking to annul this value at the moment of landing. This hypothesis is supported by the observation of behavior previously unexplained among insects, such as the reduction of their altitude by facing into winds (they

Figure 7.1
The captive flying robot, Octave, inspired by the fly. (© CNRS Photothèque, Nicolas Franceschini, UMR6152—Mouvement et Perception—Marseille.)

are braking, and the ground appears to go by more slowly), and the use of a rear wind to increase altitude (they are accelerating, and the ground appears to be going by faster).

These researchers have tested the principle of a regulator of optic flow on several robots, including a captive flying robot attached to a perch linked to a central pole, an arrangement that allows it to fly in a circle while rising and descending. The Optic-flow Control sysTem for Aerospace Vehicles (OCTAVE) is equipped with a patented optoelectronic sensor closely modeled on the neurons that detect movement. This sensor permanently looks at the ground and transmits optic-flow values to a controller that adjusts instantaneously the speed of the engine as a function of its altitude. The same principle might be used to detect and avoid lateral obstacles, and then this robot would be able to reproduce those aspects characteristic of the flight of the insect during takeoff, cruising flight, and landing—without employing heavy equipment onerous and costly in energy, such as altimeters, radar, sonar, and geographical positioning systems (GPS). The manufacturers of helicopters and drones of all kinds are already in negotiations with the inventors.

The Desert Ant and the Vikings

Ants in the forest return to their nest by trusting their sense of smell. While traveling in search of food, they deposit pheromones on the ground, the olfactory Ariadne's thread that they have only to follow to return from their expeditions.

The ant *Cataglyphis* cannot use this same means of navigation, for it lives in the Sahara Desert. The strong rays of the sun and the winds that shift the sands would soon evaporate and erase any trace of this chemical substance. Yet this ant must have an effective navigation system because when it goes exploring in zigzag fashion to find food, it manages to come back almost in a straight line to its nest. The world record is 592 meters outbound and 140 meters return! British and French ethologists have shown that the ant navigates by dead reckoning, like the Vikings of the first millennium, and that it reckons both the length of the path since its point of departure and the direction it has followed.

To reckon distance, the ant (like the fly) uses optic flow as well as data that its proprioceptive sensors supply on the itinerary that it has covered. To judge direction, it uses embedded solar compasses, visual cells that are capable of detecting in what direction the light is polarized—information that allows it to determine its position in relation to the sun, even when the atmosphere is cloudy.[2] This mechanism is very simple to put into effect and does not require either memory or complex calculations—just what suits *Cataglyphis*'s one-cubic-millimeter brain.

Robot designers at Zurich University have copied this navigation mechanism in a mobile robot in order to find a means of replacing the classic instruments of navigation, especially the GPS, when these cannot function or else break down. In the same desert where the ant *Cataglyphis* lives, they tested Sahabot 1, which uses only a solar compass based on polarized light sensors that inform it of its orientation, and Sahabot 2, which is also furnished with a mechanism of path integration—constantly informing it of the distance covered. Comparing the performances of these two robots, the engineers have shown that the simultaneous use of the two types of information generates robust navigation capacities comparable to those of the desert ant.

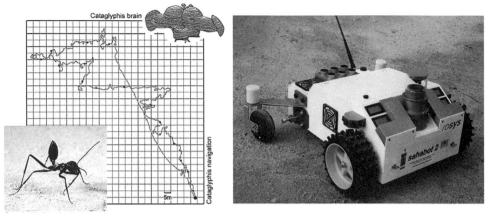

Figure 7.2
Orienting like the desert ant. Left: The ant *Cataglyphis* and its zigzag outgoing path and almost straight return. The ant's one-cubic-millimeter brain is shown at the upper right. (© Rüdiger Wehner, Zürich University.) Right: The robot Sahabot 2 navigates with the same process as the desert ant. (© Ralf Moeller, Bielefeld University.)

Hearing

Legs for Hearing

Visual organs are useful aids for moving or locating oneself, provided the milieu is sufficiently open. When couples among crickets mate, they cannot generally use vision to find each other because of the high grass that surrounds them. This is why they are attracted by *phonotaxy*, guiding themselves by sound signals. Among crickets, it is the female that finds the male. The latter chirps by rubbing its wing cases against each other, which produces a sound with the frequency and rhythm characteristic of its species. It has a repertoire of three songs—to call to combat, to court the female, and to reproduce. The female recognizes with precision the male's song by the vibrations in the tympanum found in both of her anterior tibias. The sound is conducted to these membranes by the respiratory mechanism, a trachea linked to lateral stigmata on each segment of her body. A mechanism strengthens the amplitude of vibrations on the side from whence the sound comes, thus stimulating more strongly the tympanum located on that side.

A team of neurophysiologists is closely collaborating with roboticists of Barbara Webb's team at the University of Edinburgh to exploit their exper-

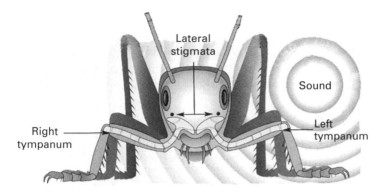

Figure 7.3
The female cricket's auditory apparatus. A sound coming more strongly from one side will unleash a movement of the legs that orients the cricket to that side. (© Dunod.)

tise on these insects' sensorimotor system. A hexapod robot, representing the female, has to find its male (a box with little biomimetic appeal) placed in an area of 70 square meters and emitting the song of a cricket. Because the experiment is performed outdoors, the ambient noise blurs the sound signal. The robot is equipped in front with two miniature microphones 1.8 centimeters apart and situated at 10 centimeters above the ground. The sound waves are processed by electronic circuits in a way similar to that of the insects. A control architecture reproducing the presumed sensory-motor circuits of the cricket ensures correspondence between these data and the movements of the legs that enable the robot to point toward the sound source.

According to some difficulties, the model of these circuits has had to be modified, suggesting to biologists that the real animal's mechanisms are not more complex, but on the contrary more simple than they had supposed. A first model suggested that a single mechanism verifies that the song is emitted at the right frequency and unleashes the approach behavior. A second model showed that the role played by the auditory peripheral system in the detection of "syllables" of the song might simplify the mechanism previously found to recognize the song. A third model, reproducing the delays observed before activating the involved neurons, demonstrated that certain neurons that were thought to be important in recognizing the song actually played no direct role.

Submarine Ears

We mentioned earlier how the communication signals between dolphins have inspired research on the transmission of subaquatic messages. The robotic dolphin "Rodolph" at Yale University is inspired by the mode of echolocation by ultrasonic "clicks" that these animals use to detect obstacles and identify objects. Rodolph is equipped with three transducers that both emit and receive signals, similar to the automatic focus on cameras. The central transducer is placed at the end of a sort of arm and emits clicks toward an object, while two other transducers, placed on sorts of rotating ears, capture the returned echoes. The effectiveness of this mode of reconnaissance resides in the fact that the ears—like those of bats—orient themselves so as to amplify the echoes as much as possible. This orientation helps to position the robot so that it turns around the object at a constant distance—like dolphins do—and thus reduces the complexity of identifying this object. In this fashion, the robot can memorize the type of echo corresponding to each type of object encountered, which permits it to recognize those it has already scanned. This study demonstrated that the exploration of a new environment with such sonar is much more reliable, much easier, and less costly in calculation than when done with a camera.

Smelling

The Lobster Tactic

The lobster identifies its prey by odor. In the day, it hides in recesses and crevices to protect itself from strong turbulence in coastal regions and from its main predators, such as the labroid, cod, octopus, crab, and humans. So it goes out at night to find its food—small crustaceans and fish—in darkness that is sometimes total. To detect its prey, it adopts a curious behavior: while periodically agitating its head up and down, it finds the flow of a few molecules that its distant prey allows to escape. Its olfactory sensitivity is great: one lobster supposedly was able to detect 33 milligrams of amino acid poured at the other end of an Olympic-size pool to which a hundred tons of salt had been added to simulate the composition of seawater! A lobster's task is all the more arduous because the turbulence of its aquatic milieu does not allow it to follow the average flow of molecules. So it must constantly reposition itself in the odorous current so as not to lose the trace of its prey. It is by *chemotropotaxy* that the lobster performs this feat of prowess—that is to say, by making comparisons with chemical concentrations detected at different places by olfactory organs situated on its antennae and under its legs. In performing an abrupt movement upward with its head, it imprisons molecules in the hairs of its antennae. It instantly compares their concentration with that of molecules that it has captured with its legs and then follows the current where the molecules are the most concentrated. By lowering its head slowly, it gets rid of the molecules captured previously, and it can resume its head thrusts upward.

Scientists at Brooklyn University, Frank Grasso and Jennifer Basil, have developed the Biologically Inspired Chemical Sensing Aquatic Autonomous Robot (BICSAAR), the size of a lobster. It does not exactly copy the morphology of its animal model, but it is equipped with two antennae on which are located sensors that detect molecules—for the needs of the experiment, not amino acids, but fluoresceine—with the same spatial and temporal resolutions as the lobster's olfactory receptors. A large median antenna sweeps the ground and simulates the sensory function of the lobster's legs. At first, by using the pendulum movement of the head described earlier, this robot did not succeed in finding the olfactory source. Researchers therefore again inspected the natural lobster's strategy; they became convinced that they had to add a second type of sensor that would enable the robot to exhibit the behavior of *rheotaxy* (a movement provoked by the rise of the current) as the animal does, but the robot does so with such residual clumsiness that one can only conclude that lobsters have not yet revealed all their hunting secrets!

Touching

Vibrissae

Instead of antennae, rodents and felines have vibrissae, whiskers that are really very sensitive sensory organs. Like the scales of sharks, vibrissae are a rat's "Swiss army knife." Rats use them for a variety of tasks: to find and distinguish food, discriminate among more or less rough textures, recognize objects, estimate the wind speed, estimate the speed of their own movement, perceive sound waves, maintain their heads above water while swimming, court a mate, orient themselves in an unknown environment, and gauge the size of an orifice before being engulfed in it or the length of a small pit before jumping across it. Numbering about thirty on each side of the muzzle, these vibrissae are equipped with three different tactile receptors that are activated thanks to muscles that enable them to move in several planes, either individually or together, synchronously or asynchronously. The vibrations of larger vibrissae help in recognizing low-pitched sounds and rough textures; vibrations of the smaller ones help recognize high-pitched sounds and smoother textures. The frequencies of vibrations may be differentiated as a function of the type of task being performed: higher for exploring, lower for finely distinguishing an object.

A robot with vibrissae is interesting to roboticists, especially for the ability to do without visual sensors that require complex data processing and that can still be the source of perceptual ambiguities—which can lead, for example, to not recognizing an object simply because the ambient luminosity has changed. Vibrissae, moreover, can direct a robot in the dark without the aid of sound waves, which might be blurred or else prove awkward in certain situations.

In the framework of the Whiskerbot project, British researchers use an electronic circuit that reproduces the sensory response of vibrissae when they brush against different textures. It appears that their vibrations are indispensable to good reconnaissance and that asynchronous and synchronous vibrations might correspond to different functions, such as the identification of an object or exploration.

Similarly, a team of Swiss researchers who are participating in the Artificial Mouse (AMouse) project have specified the influence of a number of vibrissae and the frequency of touches on the number of textures that can be discriminated. Another team in this same project has implanted in a robot both vibrissae and luminosity sensors so that it can find a source of light in a cluttered environment. The visual sensors ensure the overall direction of its trajectory, and its vibrissae enable it to follow the walls and avoid obstacles.

Multimodal Perception

The Robot Rat Psikharpax

For several years, our own team at AnimatLab, in 2007 joining the Institute of Intelligent Systems and of Robotics at the University of Pierre and Marie Curie, has been working on a European project[3] to create the robot rat Psikharpax. This wheeled robot integrates a large number of sensors and control architectures modeled on the rat, with the fundamental objective of better understanding the functioning of a vertebrate nervous system and with the technical aim of increasing future robots' decisional autonomy.

On the model of Albert Ducrocq's electronic fox, the sensory equipment of this robot's head is very varied. In particular, the data from the two sensors for peripheral and foveal vision are processed in eyes that move thanks to two small motors. Two artificial cochleas are protected by two earlobes that can pivot. An inertial platform simulating the vestibular

Figure 7.4
The 2009 prototype for the Psikharpax robot inspired by the rat. (©CNRS Pho-
tothèque/ISIR/Benoît RAJAU.) See color plate 15.

apparatus provides information on the head's linear and angular accelerations. Finally, vibrissae made of carbon fiber—thirty-three on each side of
the muzzle—serve both to discriminate textures and to recognize objects.

The work of integrating these various sensory modalities aims to understand how the brain coordinates and synthesizes the corresponding information in order to elaborate a coherent representation of the environment.
It is particularly a matter of specifying when two sensory modes should
support each other—when, for example, the fact of sniffing a baguette of
bread reinforces the hypothesis that what one is seeing is freshly baked—or
else when one mode must take over from the other because it is judged to
be more precise. It is the latter phenomenon that occurs in the "ventrilo-
quist effect," when the brain grants greater attention to vision than to
audition and spontaneously locates the sound one hears as coming from
the mouth that is seen to be moving.

8 "Wired" Control Architectures

There was a most ingenious architect who had contrived a new method for building houses, by beginning at the roof, and working downward to the foundation; which he justified to me, by the like practice of those two prudent insects, the bee and the spider.
—Jonathan Swift (1726)

The moment they concern behavior, all the examples described in the two preceding chapters call upon a *control architecture* that translates sensory data into actions (although that was not always explicitly indicated). These actions modify the robot's or animal's internal and/or external environment, and this environment sends back to sensors new information that must be processed in turn.

A control architecture is considered "wired" when the designer adjusts and "fixes" the whole organization of this equivalent to a nervous system, notably its internal parameters. This chapter shows that this type of architecture can nevertheless ensure varied behaviors and adapt itself to the environment and that it can even make behaviors *emerge* that have not been specifically programmed initially.

In contrast, "nonwired" architectures can be modified in the course of the experiment under the effect of adaptive processes such as development, learning, or evolution, and a few examples of these architectures are given in subsequent chapters.

An Artificial Cockroach

Randall Beer's dissertation, defended in 1990 at Case Western Reserve University, is important from a historic standpoint, on the one hand, because it participated in the beginning of the animat approach and, on the other, because it was one of the first applications of neuromimetic networks to the

Neural Networks—or Neuromimetic Networks

First of all, it should be noted that neural networks are described by lines of computer code and not in the form of artificial cells connected to each other. They bear this name only because there is an analogy—and a distant one—between the operations realized by the nervous system and what these artificial networks actually do.

These computer programs conceptually put to work units that communicate with each other and that are called *formal neurons* or simply *neurons*. Each of these neurons processes data that it receives from internal or external sensors or from other neurons to which it is connected, and it applies a function of activation to the result of these processes. It then assumes a value of activation that, more or less weighted, is transmitted as an input signal to the following neuron or as an output signal if the neuron is connected to an actuator. The corresponding weight is called the *synaptic weight*, which corresponds to the exciting or inhibiting effect of the connection between two neurons, the synapse, which in turn makes the next neuron a little more or a little less apt to be activated.

Neurons can differ from each other by their function of activation and by their synaptic weights. Many structures of neural networks can be used according to the types of application for which they are designed. The simplest structure, like that of a network called Perceptron, is made up of a layer of input neurons that are directly connected to a layer of output neurons. A more complex structure is made up of several layers of neurons connected to each other. Other structures implement neurons that are fully connected to each other, some playing the role of input neurons and others the role of output neurons. For a robotic application, a neural network can play the role of *controller*, the input neurons being connected to the robot's sensors, the output neurons to its actuators.

Neural networks are capable of learning through experience—which is discussed in subsequent chapters—because they relate the sensory stimulations furnished as input to the motor orders delivered as output. To do this, one can modify either the values of their activation function, or their synaptic weights, by means of different procedures. The Hebb Rule, which results in an increase in the synaptic weight of the connections between two neurons when they are simultaneously active and in proportion to their respective activation values, can be applied within the framework of a kind of learning "by association." The algorithm of *error retropropagation*, which results in modifying the synaptic weights of the inputs to a neuron as a function of the part the neuron takes in the error committed by the whole network—on condition that it is possible to compare the response that the network furnished to a given stimulus with the response that it *ought* to have given—can be applied within the framework of learning "by reinforcement." Applications of these kinds of learning are described in the next chapter. Other and more elaborate methods of learning exist. They involve networks in which the properties of neurons can be modified, not only by the electrical signals that they exchange in pairs, but also by communications simulating chemical exchanges at greater or lesser distance. Finally, the formal neurons briefly described here can be replaced by entities whose functioning is closer to that of real neurons. In particular, a growing number of applications now apply "spiking neurons," which are capable of transmitting signals similar to "natural" action potentials.

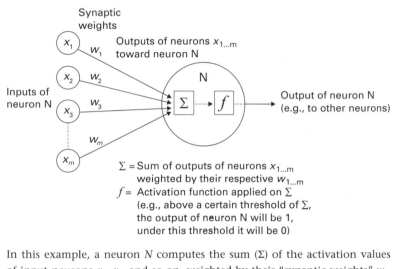

In this example, a neuron N computes the sum (Σ) of the activation values of input neurons x_1, x_2, and so on, weighted by their "synaptic weights" w_1, w_2, and so on. Neuron N applies an activation function (f) to this sum. The output value can be supplied as an input to other neurons. (© Dunod.)

behavioral control of an artificial creature. His research was conducted in close collaboration with specialists in the biology of the *Periplaneta americana*, the American cockroach.

The work of Randall Beer used a simulated cockroach, capable of moving its six legs in a coordinated way to move about in its environment looking for odorous food while avoiding obstacles. It detected both obstacles and food thanks to olfactory and tactile sensors associated with its two antennae. Food could be recognized as edible thanks to tactile and chemical sensors situated in the mandibles and could then be consumed thanks to appropriate masticator movements. Each of its actions was controlled by subnetworks of neurons, and all these subnetworks were interconnected to make up an overall system—the control architecture, properly speaking—in such a way as to ensure coherent behavior. Most of the work bore on the refinement of this network so that the animat could "survive" by not exhausting its energy reserves, the level of which it gauged by an internal sensor.

Appetitive and Consummatory Behaviors

Ethologists refer to behavior linked to the active search for food or for a partner as appetitive behaviors and to feeding and reproduction properly speaking as consummatory behaviors.

In its active search for resources, *Periplaneta computatrix* first had to move about to be able to explore the environment. Six identical subnetworks controlled the movements of each leg, so that the animal posed the corresponding "foot" on the ground, threw the leg toward the rear, then brought it forward, before posing the foot again on the ground. The robot cockroach coordinated itself with the aid of a small supplementary network that ensured the movements of the six legs according to a tripod rhythm, the most efficient way among insects: at any moment, the forward and rear legs situated on the same side were resting on the ground, along with the middle leg of the opposite side. When the animat had to turn, neurons preset to adjust rhythm activated a little more the leg situated on one side only. When it had to stop, these neurons became silent.

Other subnetworks controlled the avoidance of obstacles and the following of a food odor by interacting with the antennae's tactile and chemical sensors. Still other subnetworks controlled the recognition of food thanks to sensors in the mandibles and controlled the execution of mastication.

The Subsumption Architecture

Yet the task remained to coordinate these various subnetworks so as to avoid the execution of all the corresponding behaviors at the same time. For this purpose, the control architecture of *Periplaneta computatrix* was organized according to a hierarchy: when one subnetwork was active, it tended to inhibit all those that were situated lower down the hierarchy. It was said to "subsume" those others.

Rodney Brooks, who made the animat approach take off, made himself the ardent theoretician and defender of such an architecture, which he called the *subsumption architecture*. To demonstrate its adaptive capacities, he used it on an impressive series of robots that were quite different in their forms, their sensory-motor equipment, and their functions—especially the robot Genghis, already mentioned.

When the robot cockroach equipped with such a "nervous system" was occupied exploring its environment randomly and one of its antennae detected an obstacle, the random movement was subsumed by a behavior of obstacle avoidance. Similarly, if the animat perceived the odor of food, it continued its current activity *if* an internal sensor did not indicate a lack of energy. But if it was hungry *and* perceived the odor of food, then it directed itself toward the source of this odor by chemiotaxy. Then when it had reached the odorous source, *if* it was hungry *and if* it realized that this source appeared edible by odor and touch, then it stopped moving about and triggered mastication behavior of eating and the recharging of its energy reserves. If one or another of these conditions was not fulfilled, then it did not cease its current behavior and continued doing whatever it was doing, whether following the stream of the odor, avoiding an obstacle, or exploring.

The control architecture of *Periplaneta computatrix* represented a fortunate transition between the architectures of cybernetic robots of the past and the nonwired architectures of present and future animats. Although being entirely fixed by its conceiver, it has ensured the adaptation and "survival" of an artificial system by combining the management of its reflexes with the existence of a *motivational system*, a fluctuating state that determines at any moment the animat's tendency to execute a behavior. In effect, according to whether the simulated cockroach was motivated to feed or not—which depended on the level of its energy reserves—it did not execute the same behavior in *identical external conditions*. By this principle, a higher degree of autonomy was achieved in an artificial system.

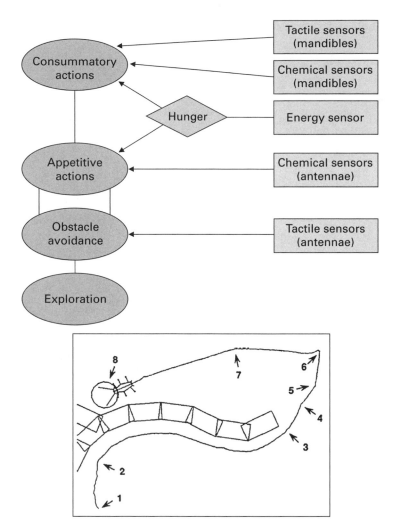

Figure 8.1

Control architecture and behavior of *Periplaneta computatrix*, Randall Beer's artificial cockroach. Top: The simulated cockroach's subsumption architecture. Each subnetwork (*ovals on the left*) is able to control the animat if it is excited by one or more sensors (*rectangles on the right*), on condition that a module of a higher level does not force it to remain inactive. Bottom: (1) By default, the cockroach explores the environment randomly. (2) It smells nearby food (situated at [8]) and, because it is hungry, the behavior of moving toward the food takes over from exploration. (2–3) Avoiding an obstacle interacts with movement toward the food. The cockroach goes along the wall, but doing so takes it away from the food, and it ends up no longer perceiving the odor. (4–6) Again, it explores randomly while avoiding the wall. (7) By chance, it finds itself in a zone where it again smells the food. It moves toward this goal, until reaching it in (8). (© Dunod.)

It should be noted that in relation to the strict hierarchy advocated by Brooks, Beer had to modify somewhat the principle of subsumption at the level of interaction between certain subnetworks, for it can happen that two controllers function at the same time. With regard both to the architecture conceived for this pioneering thesis and to those architectures still being developed today, it appears that improving on them is still largely an empirical matter that relates just as much to art as to science.

Swarm Intelligence

For the coordination of simple reflexes with a view to realizing coherent overall behavior, one naturally thinks of insect societies. It happens that certain researchers have considered these societies as a single vast nervous system in which individuals are like neurons interacting to develop a common effort. This is what they call *swarm intelligence*.

This expression covers all research that aims to reproduce and explain the collective and apparently organized phenomena exhibited by social insects, for example. The latter, although endowed with a limited quantity of cerebral matter, manage to realize constructions as complex as a beehive, a termite nest, or an ant nest and to do so without a supervisor that coordinates each individual's actions on the basis of an overall vision of the progress of the collective task. Likewise, one might wish that an artificial *multiagent* system composed of simple entities, each performing simple actions, would prove capable of producing complex collective behavior. Such a system is called *decentralized* because its global coordination is ensured only by the interactions between the agents that constitute it, and each of these agents has only a limited view of the task to be accomplished.

The application of principles of swarm intelligence is important in robotics. In place of large, expensive, and complex robots, many small and simple robots can perform the same tasks just as effectively. Although the least breakdown is likely to put an end to a single robot, the failure of one or several of them participating in a multiagent system does not necessarily compromise the success of the collective task.

Emergent Properties
To create such a system, should one minutely program each individual by thinking of the overall behavior that members should collectively exhibit? Not always. Sometimes this behavior does not arise intuitively from the initial function assigned to each robot. This is perfectly illustrated by the experiment with the Didabots, conceived for educational purposes at Zurich University.

This experiment used a set of four small, wheeled robots equipped with two infrared sensor-transmitters in front, each with a limited-range beam. They were initially placed in four corners of an arena that was scattered with small blocks made of light polystyrene. These robots moved at random, sometimes pushing the small blocks. After a certain amount of time, it was noticed that the Didabots had "arranged" the blocks in small heaps. Yet if a robot designer had the intention of conceiving of robots to fulfill this function, he would no doubt have designed a control architecture that was much more complex than what the robots actually possessed.

In reality, each individual was programmed only to *avoid obstacles*, and it was a *production default* that resulted in its function of "arranging"! When one of the beams emitted by the robot detected either a block, another robot, or the walls of the enclosure, this detection triggered a reflex of avoidance consisting of reversing and making a quarter turn to the side. The sensors were placed such that there was a dead angle between them. Now when the robot moved near a block that was located in this blind spot, no beam reached the block, and so the obstacle was not detected. The robot pursued its route and pushed the block until it approached a wall, another robot, or another block. This led it inevitably to perceive one or the other of these obstacles and thus triggered its avoidance reflex.

An outside observer might interpret this behavior by the robot as a "decision" to go look for a new block with the "intention" of placing it alongside the other, but this function of arranging was by no means intentional. It was only a consequence of each robot's particular morphology, and it disappeared if the dead angle between the two sensors was eliminated, without there being a need to change anything else in the robot's control mechanism.

In this experiment, the Didabots did not exchange any information and perceived each other only as possible obstacles. The arranging could have been performed by a single one of them, and the fact that there were four only accelerated the process. A greater number of robots would not necessarily have facilitated their task because of the time that they would have lost in avoiding each other.

When this type of "elaborate" property—such as the arranging behavior here—is combined with expected "basic" properties—such as avoiding obstacles—it at first surprises those who observe it; when it ends up being explained, it constitutes a so-called *emergent* property. Appearing at any moment and in any application in autonomous robotics, it illustrates the difficulty of understanding and mastering the sometimes complex relations between structure and function. The whole art underlying the empiri-

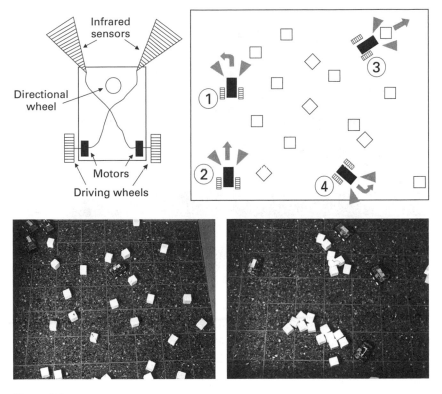

Figure 8.2

Didabots. Top: A Didabot and its environment. (1) The right sensor detects an obstacle (a cube), the left motor slows down, and the robot turns to the left. (2) No obstacle in front of the robot: it moves ahead at maximum speed. (3) Although the cube is in front of the robot, the robot does not perceive it, continues to move ahead, and pushes it. (4) The right sensor has found an obstacle (the wall), the left motor slows down, and the robot turns to the left. (© Dunod.) Bottom: The start and end of the real experiment with the Didabots made of Lego. (© Rolf Pfeifer, Zürich University.)

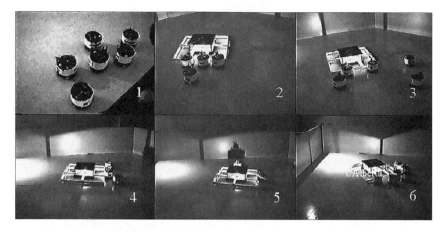

Figure 8.3
Six robots cooperating in a task of collective transport. (© Ronald Kube, University of Alberta.)

cal approach mentioned earlier consists of judiciously combining these emergent properties with the rational choices that the implementation of expected properties necessitates.

Collective Transport
Here again it is a matter of pushing a block, but one that is too heavy for a single individual to manage by itself. The few simple rules with which the little robots are equipped by Ronald Kube and Hong Zhang, at the University of Alberta, enable the animats to perform a collective transport.

Each robot has obstacle detectors and photosensitive sensors that inform it both about locating the object it must push because the latter is luminous and about the place where it should be moved, which is also associated with a light source. The program controlling its behavior is made up of these simple rules:

• Move randomly by avoiding obstacles;
• When the block to be pushed is detected, go toward it by avoiding other obstacles;
• Upon contact with the block, verify if it is located between the robot and the goal. If yes, try to push it. If the block cannot be budged or if the pushing angle is not correct, then move again at random.

Under these conditions, the robots move at random and look for contact with the block while avoiding each other and avoiding the walls of the experimental arena. They occasionally enter into contact with the block but can move it only if there are enough of them to push it in the right direction. Although the moving of the block is episodic and erratic, the robots always succeed in pushing the block to the goal without any direct communication between themselves.

Yet we note that like the programs of strong artificial intelligence, these rules rely on many implicit pieces of knowledge—such as the one that allows them to verify the alignment of robot-box-goal. Nevertheless, the principal objective here is to discern which simple reactive mechanisms, if applied collectively, are capable of replacing the controller of a more complex entity.

Coordinated Moving

Analogous mechanisms might govern the movements of great assemblies of animals. A school of fish, a herd of bison, a flock of starlings moving in coordinated fashion all seem to behave as if they were one and the same organism. These groups are adaptive because they strongly diminish the risk of predation incurred by the animals that participate in them. Ethologists think that rules can be discerned—just as Martians who are astonished that so many car drivers pass by each other without (too many) collisions in the Place de la Concorde in Paris might end up suspecting the existence of traffic rules.

Several years ago Craig Reynolds, a researcher in artificial intelligence at MIT, found out how to simulate such complex collective phenomena in a simple way. Like car drivers, his artificial creatures, the Boids, follow appropriate behavioral rules. These rules are very simple, few in number (three), and solely based on what the Boids perceive in their limited visual field. Indeed, Boids are supposed to

• Remain separate from their perceived fellows or from an obstacle by a minimal distance (rule of separation);
• Align themselves on the median orientation of perceived fellows (rule of alignment);
• Approach the median position of perceived fellows (rule of cohesion).

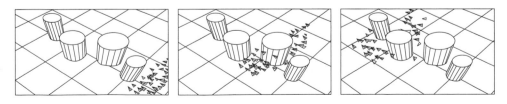

Figure 8.4
The Boids, artificial birds whose coordinated flight emerges from simple rules. (© Dunod.)

When followed simultaneously by each of the group's participants, these rules enable them to reproduce the coherence, the abrupt changes of direction, and the gathering together after avoiding an obstacle or a predator that characterize a school of fish, for example. Many other collective phenomena—the flight of birds, the herding of buffalo—can be reproduced with these same simple rules. They have even been used to simulate group effects in various animation films—such as the flights of bats in *Batman Returns* or the stampede of wild animals in *The Lion King*.

These rules have also been applied successfully to real robots. They enable a roboticist to concentrate on the trajectory that the "lead" robot will take, with the others always remaining grouped around it and following its movements.

Beyond Reflexes

Among most living organisms, the nervous system's translation of sensory information into motor orders is not always reduced to a simple reflex—that is, the same stimulus always involving the same response. Because of the work by psychologists Clark Hull and Edward Tolman, we know that the internal state can influence the subject's responses in diverse ways. Motivations, moods, emotions, beliefs, memory of past experiences—all are factors than make this state fluctuate and contribute to making the animal or the robot still more adaptive.

If it is evident that a reflex allows the rapid reaction to present circumstances, like catching oneself before falling into a ditch, the memory of past experiences can enable one to remember, for example, that both a ditch and a restaurant had previously been encountered there. In the same way, the capacity to plan and to anticipate offers the possibility of taking into account the future consequences of one's actions, such as choosing the shortest route to reach the restaurant while avoiding the ditch.

Numerous systems equipped with reflexes have already been described here, reflexes that were "wired" by those who conceived these systems. The control architecture of *Periplaneta computatrix*, as we saw, was conceived to manage both reflexes and motivations and could thus react to the animat's internal and external perceptions. In the following chapters, we describe other creations in which the capacity to memorize or to anticipate is spontaneously discovered or improved, thanks to processes of learning, development, and evolution—all of which were inspired by those same processes at work in nature.

9 Robotic Learning

Truly I came to guide you toward nature and all her children, to attach her to your service and make her your servant.
—Francis Bacon (1626)

Learning is the process by which new information is used to modify succeeding responses. Since the time of the protorobots used to test behaviorists Thorndike and Watson's hypotheses, many methods of artificial learning have been conceived. Especially used are neural networks that have allowed considerable recent advances in this realm.

The three major types of learning used in bioinspired robotics—learning through *reinforcement*, learning through *imitation*, learning through *association*—are illustrated here by a few examples, although certain systems that we previously described have already made use of them. A fourth type of learning, used instead by the GOFAI systems, is not included in these descriptions. It is *supervised* learning, in the course of which the functioning of the system can be corrected by an instructor indicating what the correct functioning might be at any moment. This method is little used by advocates of the animat approach because the natural environment rarely gives such precise feedback. When an animal runs up against an obstacle, this contact is translated into a painful stimulus, but nothing indicates to it which particular behavior it ought to adopt in order to avoid this result. In most cases, the environment merely provides a vaguely positive or negative feedback or even no feedback at all.

Learning through Reinforcement

This mode of learning exploits the principle of the carrot and the stick, or, more exactly, the laws of behaviorist learning already implemented in the protorobots mentioned previously. To the extent that certain actions by an animat are "rewarded" and others are "punished," this type of learning consists of modifying the parameters of its control architecture and hence its behavior such that in the course of time its actions are more and more often rewarded and less and less often punished.

Brachiation

Such learning was put to work on the Multi-Locomotion Robot III (MLR III), conceived by a team at the University of Nagoya, led by Professor Toshio Fukuda. This robot learned by trial and error not only to move about by *brachiation*—by swinging from branch to branch, like a gibbon— but also to walk either on four feet or on two feet, like a gorilla. Choosing the mode of locomotion most suitable to the environment or to the required task can optimize a robot's mobility. MLR III was the synthesis of Brachiator III, the gibbon, and Gorilla Robot III, the gorilla. A meter high, with arms longer than its legs, it more or less respected these primates' morphological characteristics. It was endowed with two cameras in place of eyes and grippers at the end of its arms. Its arms had five degrees of freedom, its legs had six, and its hips two.

It would have been impossible for a human to parameterize correctly "by hand" the robot's control architecture to implement directly all the kinds of sensory-motor coordination required for the task of multilocomotion. Therefore, it was by experience that this robot learned these different modes of locomotion, one after the other, in order to be able then to select advisedly which to use at any given moment.

If control over quadruped and biped walking is not an easy task, then that of brachiation is even less easy. The robot learned this mode of moving by suspending itself from the bars of a ladder placed horizontally at a certain height. The exercise consisted of swinging the lower body sufficiently to take off for the next bar, which the robot caught with one of its grippers after having let go of the bar that it held with the other. Thus, learning consisted of finding the best amplitude of swinging, the correct cyclical movements of the arms, and then the best moments for the bars to be let go and grasped; meanwhile, the only guide was the data furnished by the cameras about the distance between the bar and the gripper. The goal was to minimize this distance. Any movement engendering greater distance was "punished," whereas those movements leading to smaller distances were "rewarded."

Duly surveyed by its creators (who caught this costly prototype each time it missed a bar!), the robot succeeded in accomplishing this difficult

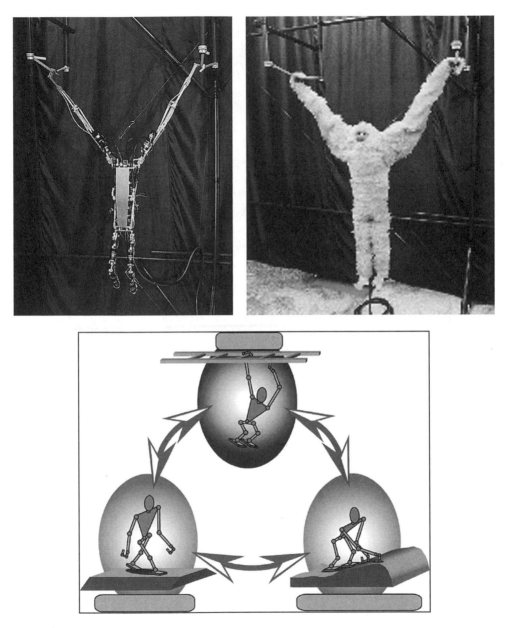

Figure 9.1
Brachiator III and Multi-Locomotion Robot III. Top: The robot Brachiator II (with
and without its fur covering), which learned to move from bar to bar. Bottom: The
multilocomotion of MLR III allows it to choose between biped, brachiation, and
quadruped locomotion. (© Toshio Fukuda, Nagoya University.)

exercise and in moving the length of the ladder without falling, even when the bars were irregularly spaced.

The two other modes of locomotion having been learned in the same way, the next stage of the project was to teach the robot how to choose the one best adapted to the nature of the environment in which it moved so that it could economize the energy it was spending.

And Psikharpax?

In our laboratory, with the goal of enriching the capacities of the robot rat Psikharpax, a robot is also currently learning by trial and error a task that seems apparently easier than the one just described. Like countless generations of laboratory rats, it has to find in a cross maze the lighted end of a branch that signifies it will find a reward there—in this case, drops of "water" delivered by a fictional distributor. One supposes that the robot rat, like its natural model, is "thirsty" before the experiment begins and hence motivated to find the correct spot. As soon as the robot rat reaches its goal, the end of another branch is randomly chosen to be lit up, and another trial begins. The robot thus has to learn to go to the center of the maze, to go to the end of the illuminated corridor, then to "drink" when it arrives at the end. This is an example of stimulus-response learning well known to experimentalists, which makes the robot choose, as a function of both its motivation and what it sees at any moment, the right action to perform—such as to go straight ahead, to turn, to stop, to drink. The robot is rewarded only when it reaches the distributor.

But this task is not as simple as it appears. To learn these behavioral sequences, the robot must favor the correct succession of actions to execute in order to obtain its reward. The problem is that the delivery of the reward occurs only after a delay because it won't be received until the rat has reached its goal, far from its point of departure. Once at the goal, the rat must try to "remember" which decisive actions, among all those it might have performed in this environment, led to this result. Is it the one it performed just near the goal or the one a little farther back or the one at its point of departure? Its memory capacity does not allow it to retain all the sequences of actions that led it from any place in the maze toward the reservoir. So it must apply a method that improves its behavior in an incremental fashion without its being obliged to retain all its past actions.

Among the solutions that computer experts envisage for this type of problem is an architecture called "actor-critic," which is composed of two modules: as a function of current sensory information, the *actor* decides on the action to perform, while the *critic* evaluates the reward that the

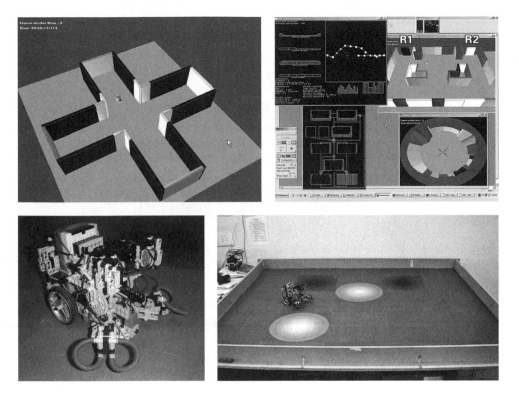

Figure 9.2

Psikharpax's mazes and "avatars." Top left: The virtual cross maze in which Psikharpax (represented here in the form of a commercial "avatar" robot) performs its stimulus-response learning. The reward is at the end of the "lit" branch, to the robot's left, and changes randomly at each attempt. Top right: When Psikharpax looks for two types of resources (R1 and R2) in an environment, the experimenters can monitor its view of the situation (lower top right), the way it builds its spatial representation (upper top left), and the way it selects its actions, taking into account its motivations (lower top left). Psikharpax chooses to go to R1 or to R2 when its motivations push it toward one or the other. (© Loïc Lachèze and Mehdi Khamassi, AnimatLab, LIP6 & ISIR.) Bottom: A Lego robot (another "avatar") actually performed this last experiment with only two light sensors. Black and white zones correspond respectively to the two types of resources R1 and R2. (© Benoît Girard and Vincent Cuzin, AnimatLab, LIP6.)

present choice will end up entailing in the future. In these conditions, the actor chooses at any moment the action that the critic considers the most promising, and thus learning consists of comparing the expected reward with the reward that is actually obtained, such that the difference between them diminishes over time.

If such an architecture interests animat designers, this is because it does not require memorizing the sequence of all the actions successively performed, and because it strangely resembles the way in which mammals, especially the rat, seem to learn a similar task. Thus, within nervous structures called basal ganglia, the roles of the actor and the critic might be held by two distinct parts of the striatum, one of the components of these ganglia.

In effect, neurophysiologists have discovered that one of the two parts—the critic—controls neurons that secrete a particular neurotransmitter, *dopamine*, that acts on the other part—the actor. Neurophysiological recordings have shown that these neurons act as if they were taking account of an error in predicting the attribution of a reward. If they were initially activated by the arrival of an unexpected reward, they will discharge less and less every time this recompense is delivered again in the same conditions. However, if the animal is expecting a reward and it does not happen, the activities of these dopaminergic neurons are at a level much lower than their basic level because the error of prediction was "inverse." The actor-critic model conceived by computer programmers seems in fact to have been invented by evolution about 230 million years ago!

The functioning of this type of neurons and the organization of basal ganglia are behind the control architecture of a virtual Psikharpax that succeeds in the task of the cross maze after trials as laborious—but ultimately as effective—as those that can be observed among real rats. In the course of successive learning trials, the robot begins to be "surprised" by the delivery of a reward when, in the presence of the goal, it decides for the first time, by chance, to "drink." The emission of dopamine toward the actor part of its "striatum" is going to modify certain synaptic weights, with the consequence of augmenting the chances that it will in future choose systematically the same action in the same circumstances. Then the motion that it performed just before drinking (moving straight toward the reservoir) will from then on be reinforced by the same procedure. It will be the same for all the "correct" actions that preceded the latter one and that will finally lead the robot in an effective way, from several points of departure, to the place where it will be rewarded. There is no need for the robot to "seek to remember" at any instant which one action should follow another—all is now inscribed on its modified synaptic weights.

The same architecture inspired by these nervous structures also enables a real robot to select its actions as a function of two motivations that its creators gave it: the need to "ingest" energy in particular sites and the need to "digest" it in other places so as to make it usable. According to the instantaneous importance of each of its motivations, the robot is capable of deciding what to do at any moment in order to reach autonomously one or other of these sites without a human's aid. Contrary to other mechanisms of selecting actions that are less biomimetic, this control does not entail any "behavioral dithering" whatever—the syndrome that killed Buridan's ass when it hesitated so much between two equally desirable actions (to eat hay or to drink water) that it died for want of deciding.

Learning through Imitation

Only the so-called higher primates, such as great apes and humans, are capable of imitation. In 1990, a team of Italian neurobiologists directed by Giacomo Rizzolatti put this characteristic into relation with the activity of specialized neurons, the *mirror neurons*. Located principally in the frontal cortex, these neurons seem to constitute an interface between observation and action, between perceived movements and executed movements. In fact, it has been shown that these neurons are activated when the animals execute a certain movement, but also when they perceive this same movement executed by another or even when they merely anticipate it.

This discovery[1] implies that a natural or artificial system can extract information about the movements that it perceives without really practicing them and then reproduce them because the zones of preparation for these movements are activated. These principles have been exploited so that a human can "show" a robot new actions to be executed without its actually needing to practice them endlessly. In a very simplified way, learning happens as follows. First, the robot must be provided with a repertoire of motor primitives associated with perceptions. For example, seeing a human arm draw a curve is associated with this gesture, seeing a human arm draw a straight line is associated with another gesture, and so on. The human then executes a movement composed of several of these primitives in front of the robot—by tracing a capital B, for example. The robot must then recognize, within the flow of its perceptions, the gestures with which it has already associated some primitives and then learn to associate them appropriately so that it can reproduce the figure executed by the human.

In this way, to achieve a B it will draw a vertical line, then add a curve from the top to the middle, then another similar curve—rather than draw an A or make a hand wave. Moreover, it can improve its imitation by slightly modifying its primitives.

Many studies bear on learning through imitation among robots—inspired closely or tangentially by neurophysiological findings—notably the one by Aude Billard, a researcher in Lausanne and one of the first to explore this domain. Robota, the doll Billard developed, is 60 centimeters high and equipped with visual and auditory sensors. Its mouth can move, its arms are articulated, its hands are prehensile. It is capable of imitating simple gestures performed in front of it by a human, such as lifting an arm or turning the head or even both at the same time. It is also capable of relating the sounds it hears to the means of pronouncing them, and so it can emit simple sentences—although still a little choppy—such as "My na-me is Ro-bo-ta."

The Robota doll, apart from its contribution to the progress of knowledge about learning through imitation in artificial systems, is involved in a program of education called AuRoRa (Autonomous mobile Robot as a Remedial tool for Autistic children). These children have a deficit in their capacity to imitate that is related to their difficulty in communicating, which in turn may be due to a dysfunction in their mirror neurons. Corresponding research aims to engage these children in a game that allows them to learn to imitate in the hope of improving their faculties of social integration; one approach attempts to lead these children to perceive, by means of very predictable behavior, that the doll facilitating this task is imitating *them*. This program does not aim to cure autism, but rather to specify certain characteristics that are difficult to discern with any other method.

Learning through Association

The third mode of learning is called *unsupervised* and concerns situations where the environment gives the system no information whatever on the value of actions it has performed or on the knowledge it has acquired. This is how one learns that shrubs are smaller than trees, that contrary to many birds the ostrich does not fly, and that when one enters a kitchen one is likely to see a fridge. The association of certain characteristics takes place according to lived experience, and no punishment or reward is associated with the results thereby obtained.

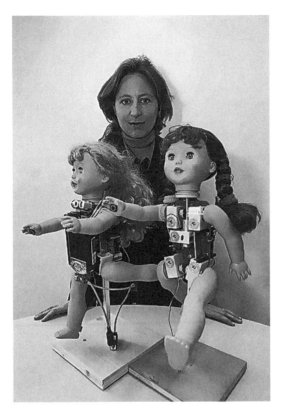

Figure 9.3
Aude Billard and two prototypes of the Robota doll. (© Alain Herzog, École poly-technique fédérale de Lausanne.) See color plate 10.

Categorizing Objects

This mode is how a little robot with wheels, equipped with a panoramic camera and a gripper, was able to learn to recognize two types of objects in an environment (cylinders with small or large diameters). Of course, a human would have immediately been able to tell them apart by their different sizes. Instead, the robot, rather than using its camera, associated the objects with the number of wheel spins necessary to skirt around them—by odometry. It finally classified the objects into two categories, "short move" and "long move." Moreover, having tried several times to lift them, it could add to the first category the characteristic "can be lifted" and to the second "cannot be lifted." Now when it encounters any "long move" object, it will no longer try to grasp it with its gripper.

This learning is not so simple, for it requires many prerequisites, such as knowing when a complete trajectory around the cylinder has been completed. Nevertheless, it illustrates the fact that the representation of the world elaborated by a human, an animal, or a robot depends on its sensory-motor equipment. We have a great deal of difficulty in imagining what we would do if we had the same morphology as this robot and were equipped with the same sensors and activators because our natural tendency is to capitalize on vision, the sense that is most developed in us primates. It is for this reason that it is important to give the robot the opportunity to build its own model of the world so as to avoid the bias that our anthropocentrism is so ready to generate—a point to which we return later.

Orienting in an Unknown Place

This mode of learning by association is also used in applications where robots' localizing and navigating are inspired by neural processing in rodents.

Living in a more complex environment than that of the desert ant, rodents have at their disposal more varied mechanisms to adapt to it. For the past two decades, neurobiologists have accumulated a considerable amount of knowledge about the way in which rats (and many other mammals) locate and orient themselves in their environment. They have become convinced that these animals elaborate an internal representation of their environment—sometimes called a *mental* or *cognitive map* (first suggested by Edward Tolman back in 1930). Examining this map "mentally," rats can choose the shortest route and avoid dangerous places.

This representation relies on the functioning of specialized neurons, such as place cells,[2] head direction cells,[3] and grid cells.[4] These cells have learned to be activated by associating and integrating various sensory data (notably visual, olfactive, auditive, tactile, and proprioceptive data).

Place cells permit the rat to locate itself in the environment. Indeed, when a rat explores an unknown space, it learns that one configuration of stimuli characterizes one place—for example, what it sees, hears, smells, and touches there—and another configuration signifies another place. Connections between neurons will then be strengthened or attenuated such that certain place cells will henceforth be activated when the first configuration is recognized another time, and others when the second configuration occurs. According to which place cells are activated, the rat will always know where it is precisely situated.

As their name indicates, *head direction cells* are active only when the animal's head is oriented in a specific direction, independently of the rat's location in its environment. Each of these cells presents a maximum activity for a given direction of the head, called the direction of *preferential discharge*. In a way analogous to what is observed with place cells, head direction cells perform a complex integration of data coming from several sensory channels. For example, the activity of these cells changes when a strong visual landmark in the environment is moved or when the rat perceives through its vestibular system that the platform on which it is standing is itself being moved.

Grid cells are activated in places that are regularly distant from each other, each cell covering the environment with a sort of grid whose nodes correspond to places where the cell is activated. The mesh size of these grids varies from one cell to another. Thus, even in the absence of certain sensory data, these cells will enable the animal to locate itself correctly and estimate the distance it has covered. The most astonishing thing is that these grids adapt to environments of all shapes and sizes.

These three types of cells learn extremely rapidly to activate reliably in an unknown site; they constitute the neuronal bases of the spatial representation mentioned earlier. The rats use the map built in this way to move effectively in daylight but also at night—because olfactory, auditive, tactile, and proprioceptive kinds of information, like visual information, are integral parts of this system.

Roboticists have tried for years to take advantage of such information about localization, orientation, and distance—which do not require highly specialized sensors (such as laser telemeters) whose use is not suitable in all kinds of environments. The challenge is to realize through learning the correct associations between pertinent sensory data in such a way as to implement the three sorts of neural activations that have just been discussed. Another challenge is to integrate all these data so that they form a coherent system.

One of the first biomimetic models of navigation to meet these challenges was developed by Angelo Arleo, first at the École polytechnique fédérale de Lausanne (EPFL) and then at the University of Pierre and Marie Curie in Paris. A small mobile robot equipped with panoramic vision, with a belt of infrared sensors and an odometric system (counting wheel spins and representing proprioceptive sensors) can build a mental map of its environment, a square arena whose walls are covered with black and white stripes similar to bar codes. At one place is a light that remains lit throughout the experiment. When the robot enters this unfamiliar environment, it initializes its sense of direction relative to an absolute direction that it chooses arbitrarily. While it explores, it learns to correlate an outstanding

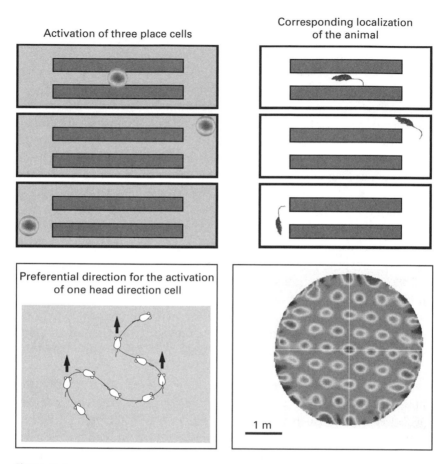

Figure 9.4
Place cells, head direction cells, and grid cells. (© Dunod.) Top: A very simple maze with two parallel partitions in the midst of a rectangular environment. As a rat moves around in this maze, certain place cells are activated in its brain, and others are not. Recording the activity of different place cells (left column) shows in which region of the maze the rat should find itself (right column) so that the cell in question is activated: dark gray indicates the zone where the cell is very active, light gray the zone where it is less so, and medium gray the zone where it is no longer active. Bottom left: Working principle of head direction cells. One of these cells is activated only when the animal has its head turned in a certain direction; another cell is activated only when the head is turned in another direction, and so forth for several possible directions. Bottom right: A given grid cell is activated (more = dark gray, less = light gray, not at all = medium gray) each time the rat moves a certain distance from the last place that triggered its activation. Another cell will activate in places that are closer to each other, and yet another cell for more distant places, and so on.

visual cue—the light source, for example—with the chosen direction. When it comes back to this environment, it will be able to reset its absolute direction with this cue and maintain a coherent representation of its orientation.

The robot now has a reference point to create its place cells, for it is going to systematically make excursions in circles from a point of departure —a behavior that is observed among real rats. During these patrols, it associates its various locations with its external and internal perceptions. It also uses them to associate consecutive locations and elaborates an internal map that is called *topological* because it links sites without being specific about the distances between them—like a map of subways on which the distances between stations are not faithfully respected. Once established, this spatial representation enables the robot to reach a resource previously memorized but not visible or to plan a detour through already explored places toward a given goal when an unexpected obstacle prevents reaching this goal more directly.

Adding grid cells would make it possible to take distances between memorized places or objects into account, to position the resources more precisely. The robot might also take shortcuts that it has never practiced— which it could not do previously because of the risk of wandering around a goal without ever reaching it.

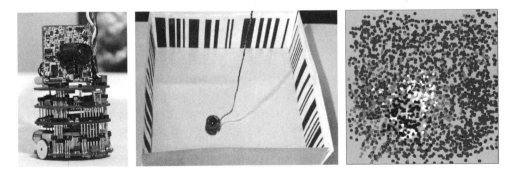

Figure 9.5

A small robot constructs its spatial representation. Left: A small robot about 6 centimeters in diameter equipped with panoramic vision. Center: The robot's experimental environment; it can use black and white bands as landmarks. Right: The robot's spatial representation. Because certain place cells are activated in its "brain" (gray levels indicate strong, medium, weak, and no activation, respectively) and others are not, the robot "knows" where it is. We notice the concordance between the zone activated in its map and the place where it really is within its environment. (© Angelo Arleo, École polytechnique fédérale de Lausanne.)

And Psikharpax?

Psikharpax uses a spatial representation and now has at its disposal several navigation strategies that it can exercise both day and night—some cognitive, with the help of a map, and others reactive, without a map. The two modes are necessary, for they prove to be complementary. In fact, although possessing a mental map offers many advantages, it can have defects: it calls on a great number of formal neurons, which might make its building and updating rather complex if the environment is large.

This is why, when a map is big and difficult to build, it is useful that Psikharpax can trust on reactive strategies, such as *taxis*, which consists of relying on a sequence of visual points (the one it learned in the cross maze, for example), or *praxis*, which consists of memorizing a sequence of movements, such as turning to the left three times in a row and then once to the right. It is also preferable that it chooses this type of strategy when the environment remains the same: it is useless to recruit the multiple neurons devoted to the spatial representation when the taxis or praxis habits are already well anchored in place.

One question that neurobiologists have much studied but been unable to resolve concerns knowing by which exact mechanisms the selection among these different strategies—whether the strategies rely on a mental map or not—is made. Our team, in collaboration with other partners, is currently developing a model to answer this question.

Collective Learning

Using both learning through reinforcement and learning through association, the robotic Talking Heads at the Sony Computer Science Laboratory in Paris collectively forged a vocabulary. The experiment was designed by Luc Steels, Frédéric Kaplan and Pierre-Yves Oudeyer to find how languages emerge, based on language games applied to agents that might be robots or software. The long-term goal was to help robots understand each other and to foster communication between robots and humans.

Two robots ("speaker" and "listener") armed with a camera are communicating about the colored shapes found on a screen. Several characteristics of these forms (size, geometry, color intensity, etc.) can be captured, as well as their arrangements (on the right, on the left, etc.). A series of syllables can be used to associate a word with one such characteristic. By being alternately "locutor" and "interlocutor," the robots observe a form and construct a simplified representation of it and can associate this representation with words constituted of the available syllables taken at

Figure 9.6
Left to right: Two robots "look at" a screen on which symbols of different shapes and colors appear in various positions. The "speaking" robot pronounces a word that it associates with one of the symbols. The "listening" robot must figure out the shape and color chosen by the other robot. (© Dunod.)

random (for example, *wa-ka-bu* or *ma-ka-ne-na*). During an interaction, the locutor gives one of the words to the interlocutor so it has to identify the designated shape. For example, if the screen exhibits a red square in the upper left, a green circle in the middle, and a blue triangle in the lower left, the locutor will say "*ma-le-wi-na*," which signifies for it "upper left red." On the basis of this clue, the interlocutor tries to guess which is the chosen shape and indicates its choice by pointing its camera at one of the objects. If it points to the right object, the locutor will use this word more in the future. In the opposite case, the interlocutor memorizes the term used. When this process is repeated many times, there emerges a vocabulary common to the two robots, a vocabulary that has no meaning for an external observer.

This experiment has been reproduced many times, especially in museums and at a distance with the Internet. It has used real and virtual robots and sometimes humans, demonstrating that a lexicon and a system of shared categories can emerge from very local interactions between agents and that this vocabulary can constantly evolve, for example, when "naive" individuals are integrated into the population.

A similar type of learning was imposed on Aibo, Sony's talking robot dog. A human pointed to a red ball and said, "Look, ball." Because there were several red objects in the dog's environment, the human showed the dog those that were not balls and started again to point to the ball from several different angles. After several trials, Aibo could answer "ball" to the question "What is that?" when someone pointed to the object. It also acquired more subtle knowledge, such as the fact that an object that someone says is "on my left" is on its *right* if it is facing its interlocutor, but indeed on its left if alongside him or her.

Figure 9.7
Aibo, its red ball, and its teacher. (© Frédéric Kaplan, Sony Computer Science Laboratory.) See color plate 12.

This experiment might recall Cyc, in which Doug Lenat and Ramanathan Guha tried to inculcate common sense into a computer and to teach it to communicate in natural language. Although the Talking Heads results appear more modest, they are no less interesting: here the meaning of a word emerges with little human a priori input. The researchers see in this prototype a means of studying how capabilities (cognitive or not) can emerge from a tabula rasa—in other words, how a system can "develop."

10 Robotic Development and Evolution

You arrive before nature with theories, and nature flings them on the ground.
—Pierre-Auguste Renoir (1925)

Development

Contrary to Athena or Dionysus, no multicellular animal has appeared "fully formed." The series of the most precocious events of its life, the organic development that begins when it is only a zygote (a fertilized egg), has an adaptive value. The construction of its anatomy and the setup of its physiology enable it to face gradually the innumerable constraints imposed by interactions with both internal and external milieus. Thus, each organism molds itself into the ecological niche in which it is going to live, and its development will be influenced by the physical and social experiences that it will have. This ecological niche will also be modified in the course of time. This is why the development of any particular organism will be different from that of a fellow creature that might have the same genes, whether it was born in the same era or sometime later. The environment's influence on the genetic inheritance brings about a morphological and behavioral variability that is the guarantee of the species' survival.

The morphological development of robots remains a roboticist's dream. Some have tackled it by taking inspiration from the evolution of species, as we will see later. The examples that follow concern only the development of control architectures.

Aibo on a Learning Mat
The Sony Computer Science Laboratory in Paris, which had already been involved in the Talking Heads experiment described in the previous chapter, conducted another very interesting program.

This program was the consequence of limitations noticed in Aibo's learning capacities: Aibo did not manage to extend to more than twenty the number of words it could associate with surrounding objects. Like an unruly student, it did not manage to concentrate its attention on the designated object unless the researchers shook this object in its field of vision. Therefore, they chose to exploit the robot's "curiosity" to enable it to choose what it "wanted" to learn.

Aibo's brain was equipped with two modules for learning, one to predict the consequences of its actions, the other to predict the quality of the former's predictions. What is called "curiosity" is the robot's propensity to choose the action that contributes the most to improving this quality. Let us imagine that Aibo can press three buttons that result in displaying on a screen either (1) a fixed X or (2) an X moving back and forth along a diagonal or (3) "snow," luminous specks randomly lit and constantly changing in position. Once it has tapped haphazardly on the three buttons, its sense of curiosity will first lead it to press the first button several times, for what is going to appear on the screen will be easily predictable, and its learning will thus make rapid progress. Then, having nothing more to learn, it will start to press the second button, but while the position of the X is more difficult to predict than previously, it can "hope" to learn something. When it has learned to predict the X's trajectory, it will press the third button, but it will quickly stop doing so, having "understood" that the random position of the lighted specks is unpredictable and that an extra learning effort will be perfectly useless.

Left to itself on a baby's learning mat, Aibo explores the sensory-motor consequences of various actions it can perform, such as moving a leg, opening its mouth, hitting a suspended object, yelping, and so on. Each time, the chosen action is the one that optimizes its learning. Like a baby, it is interested for a while in a certain behavior that it reproduces several times over, and then it suddenly abandons that and goes on to something else. Thus, it slowly elaborates its motor repertoire and learns that a suspended object will swing if given a swat or that it can be nibbled, that yelping when one of the experimenters enters elicits a yelp from him or her—a deliberate choice by the researchers to communicate with their robot. This type of learning has its disadvantages: Aibo can reckon that it is interesting to learn to bang against the walls!

Nobody tells Aibo what is interesting. It can turn toward sensory-motor associations that it has not yet understood, like the baby who rips off the wrapping paper on a present it has just got. It will quickly tire of very

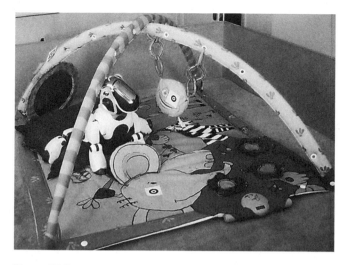

Figure 10.1

Aibo, on its learning mat, learns to develop its behavioral repertoire. (© Frédéric Kaplan, Sony Computer Science Laboratory.) See color plate 11.

complex learning, too, but it will increase the richness of its interactions with the environment as it has more experiences.

This kind of research was pursued by putting two Aibos, one of which was more experienced than the other, on the learning mat. There the more experienced Aibo's faculties of imitation accelerated the increase in the "naive" robot's motor repertoire. Moreover, a kind of primary communication between these robots emerged.

The interest of this study resides in the fact that any robot can constitute its own behavioral repertoire and forge a model of its interactions with the world. Indeed, it is the only one capable of elaborating this model because its human conceiver cannot put him or herself in its place, not having either the same body nor the same sensors nor the same actuators. This is what the biologist Jacob von Uexküll understood at the beginning of the previous century when he endowed each organism with a representation of the world it constructs by itself—its self-centered world (*Umwelt*), in which each object has its own *functional tonality* (*Funktionale Tonüng*).

The psychologist James J. Gibson later formalized this same idea differently under the term *affordance*, which refers to the fact that what one perceives is the opportunity for action that an object permits. For a human to perceive a chair is to apprehend an opportunity to sit down or to stand on it to reach an elevated object, and so on. Affordances are linked to the subject's sensory-motor equipment and are therefore specific: a fly will not distinguish among walls, ceilings, and floors because it can walk on all three surfaces, which a human obviously cannot do. Affordances are even individual: the same stairs will be perceived by some humans as practicable and by others as impracticable as a function of the length of their legs.

Stages of Development

The following research tackles a human's more specific sensory-motor development.

The robots Cog, developed at MIT in Rodney Brooks's laboratory, and Babybot, developed at Genoa by Giulio Sandini, serve to study human sensory-motor and cognitive development, taking inspiration from neurological and psychological studies such as those conducted by the Swiss psychologist Jean Piaget on a child's stages of development. These two robots do not have legs, but only a head on a torso, with two arms for Cog and a single one for Babybot. They are provided with visual, auditory, vestibular, kinesthetic, and tactile sensors. Their brains are made up of neural networks implanted in many computers functioning in parallel.

If the researchers gave both robots a vaguely humanoid aspect, they did so in order not to repeat the error of the GOFAI pioneers, who wanted to produce a robot as intelligent as humans, but rather in order to respect what one might call the *principle of embodied cognition*: to develop human sensory-motor and cognitive capacities, it is best to have a human morphology.

The development of these robots was incremental: the first stage consisted of teaching them their "body schema"—that is to say, to differentiate their body from the rest of the environment. They would then proceed to a second stage, learning to interact with objects. This would enable them to tackle the third stage, consisting of understanding the relations among objects.

According to those who conceived them, these robots have already acquired the capacities of a baby between six and eighteen months old. The body schema was learned by associating the movements of the body with the various perceptions resulting from these movements. Of course, this association entailed tackling the problems of integration mentioned earlier. For example, Cog in some way appropriated its new hand when the researchers moved its fingers because it succeeded in correlating the various data processed by its proprioceptive, visual, and tactile sensors. It was able then to learn to grasp an object, then to follow an object with its eyes, to choose one object when several objects were displayed before it,

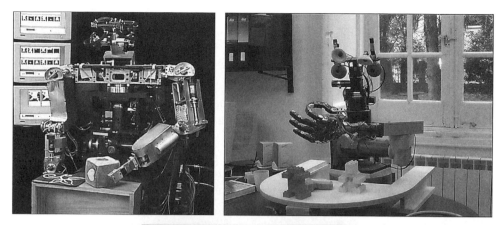

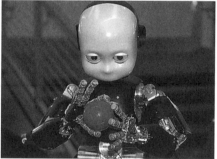

Figure 10.2
Various robotic platforms in the course of sensory-motor development. Top left: Cog. (© Rodney Brooks, MIT Computer Science and Artificial Intelligence Lab.) Top right: Babybot. (© Giulio Sandini, University of Genova.) Bottom: iCub. (© Giulio Sandini, University of Genova.) See color plate 13.

and even to look at an object pointed out by a human. For its part, Babybot learned to grasp an object, to adjust its movement according to whether the object was heavy or light, and to recognize objects already encountered, even if they were seen from a different angle. It turned its head toward a sound object and could follow it with its eyes. It learned that a spherical object can roll, unlike a cubic object, which has to be pushed in order to move it. It also learned to recognize the movement of an experimenter who was pushing a similar cube.

Teams working on these projects have already undertaken another one: constructing iCub, a robot resembling a two-year-old child that has both Cog's and Babybot's capabilities, but that also possesses posterior limbs. It will therefore be able to move, at first by crawling, then by supporting itself on four limbs, and finally on two. This learning is now under way with the collaboration of several European laboratories—including ours.

Developing Emotions

To speak of a shrew's joy or an eel's sadness seems to derive more from poetry than science. Yet many scientists agree that all vertebrates (and perhaps cephalopods) express emotional states in their own ways. Ever since Antonio Damasio's book *Descartes' Error: Emotion, Reason, and the Human Brain*,[1] it is accepted that emotions play a decisive role not only in survival and in social relations, but also in every decision taken at any moment. In short, they play an essential role in an individual's adaptation to its physical and social milieu.

In time, animats will acquire such capabilities. Some projects are now focused on this goal within the framework of the "social" development of adaptive robots. These programs are essentially designed to improve interactions between robots and humans because what most frightens future users of domestic robots is not being able to detect what "mood" they are in. To appear a little more transparent to those around oneself, to give clues regarding one's state—these are major functions of emotions.

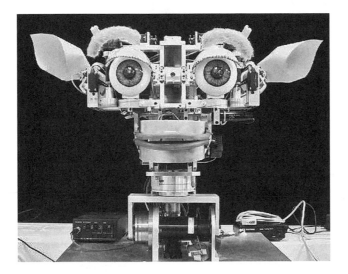

Figure 10.3
Kismet, the expressive head. Many motors animate Kismet's face, enabling it to express emotions. (© Rodney Brooks, MIT Computer Science and Artificial Intelligence Lab.) See color plate 16.

To appear to feel human emotions requires being anthropomorphic. The Sociable Machines Project at MIT directed by Cynthia Breazeal has developed Kismet, an expressive head endowed with sensory modalities such as vision, audition, and kinesthesia. It can emit sounds, turn its head and eyes, move its eyebrows and its mouth, and thus give the impression that it feels anger, disgust, fear, joy, surprise, interest, sadness, boredom, or excitement. Researchers were inspired by knowledge about childhood development in teaching the robot which emotion to express in which situation. Kismet can now follow a person with interest with its eyes when that person enters the room. It recognizes vocal intonations and as a result can change the tone of its own vocalizations. It expresses sadness when scolded and joy when praised. If someone approaches it too rapidly with an object, it expresses fear or anger. If a toy does not interest it anymore, it turns its attention away and tries to engage the person present in conversation.

Kismet's control architecture makes use of six modules that are permanently interacting, including a behavioral module that integrates external information—such as extracting simple and complex characteristics (recognizing a person or an object)—and internal information coming from a motivational module that governs emotions. The behavioral module constantly sends back its "synthesis" to all the sensors as well as to the motivational module so that they all can modify their inner workings through learning. The behavioral system simultaneously controls the commands sent to the different motors. The latter make facial expressions and give speech the tone suited to the actual situation, with more or less intensity.

Kismet's interactions with its human entourage are impressively natural, for the robot deploys in a coherent manner the appropriate emotional signs. In turn, it influences the behavior of its interlocutors, as in the case of human exchanges. Beyond this success, the scientists are pursuing a more fundamental goal. It appears in effect that this form of communication enables them to know at any moment if the robot's learning process is going well. For example, if the intensity of some expression is too strong (or not strong enough) with respect to a particular situation, the tutors can modify immediately their interactions in order to rectify the robot's corresponding expression. Thus, they find themselves in the same conditions as parents who, realizing that their child is expressing an emotion too great or too small for the context, immediately bring to bear the necessary information for it to modify its reaction.

Kismet's emotional learning has not finished. An interesting new situation for it to confront would be the revelation of the number of computers necessary for its enormous calculation needs. No doubt it would react with a quite new emotion, mingling surprise, fear, sadness, and excitement!

Emotional robots are swarming in many laboratories. For example, the Human MAchine INteraction on Emotion (HUMAINE) association, which gathers together many international teams, and the recent European project "Feelix-growing" (FEEL, Interact, eXpress: a Global appRoach to develOpment with INterdisciplinary Grounding) aim to develop human-robot interactions via the expression of emotions.

In Japan, programs of humanoid robots interacting with people are very advanced. One of the reasons for this advancement is that the aging population makes it essential to refine robots that can help old people at home. For the time being, the scientists are looking for the most perfect resemblance to a human; for example, Professor Hiroshi Ishiguro at Osaka University has conceived a clone robot of a Japanese television personality (with her permission). Repliee Q2 has a silicon skin with multiple sensors and exhibits respiratory movement, blinks its eyes, possesses a great variety of facial expressions, and moves its arms and hands rather naturally. It possesses about two thousand responses to actual questions and is able to maintain a conversation with a human. It also deploys emotional and gestural expressions adapted to its interlocutor by imitating what it perceives in him or her.

Capitalizing on this success, Ishiguro has produced his own clone robot—a gadget useful for occupying his office when he is abroad!

However, scientific interest in these diverse studies of robots' emotivity is still limited because the corresponding robots express only "surface" emotions. Antonio Damasio and many other scientists are persuaded that although robots may exhibit external behavior, they will long remain incapable of producing intimate feelings.

If emotions play a much more fundamental role among living systems, it is because they emerged in the course of evolution—at the same time anatomy and physiology did—as an innovating function benefiting an organism's adaptation. In the course of time, organisms were able to find, among the billions and billions of possible configurations of their nervous and hormonal systems, those best adjusted to their sensory-motor equipment as well as to their need within their particular ecological niche. Thus, a researcher using a commercial robot is often in the situation of a creator who is given a fully equipped animal that lacks control systems. What do you do to make the sensory and motor organs most efficiently connected

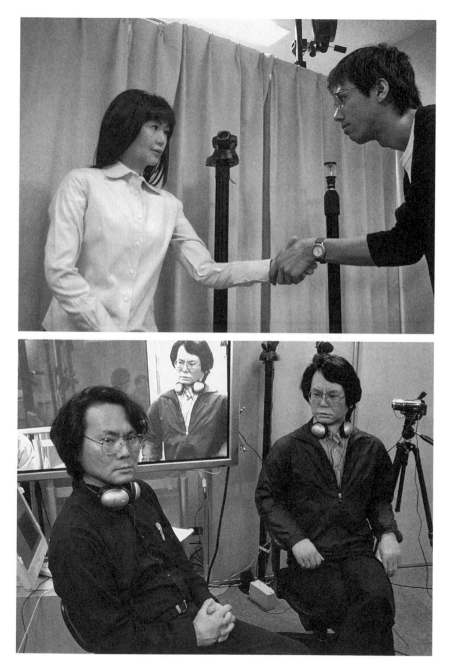

Figure 10.4
Professor Hiroshi Ishiguro's "clones." Top: Repliee Q2. (© Hiroshi Ishiguro, Osaka University and Kokoro Co., Ltd.) Bottom: Professor Ishiguro on the left and his clone Geminoïd on the right (or maybe vice versa). (© Hiroshi Ishiguro, ATR Intelligent Robotics and Communication Laboratories, Osaka.) See color plate 14.

to each other? As we have just seen, various types of learning can help adjust an architecture's functioning details in order to increase the efficiency of its sensory-motor control. Yes these methods have their limits, for a vast part of the space of possible configurations has not been explored. Therefore, evolutionist methods might help close this gap.

Evolution

Around 1960, a group of researchers led by John Holland at the University of Michigan, followed by other American and German researchers, conceived evolutionary methods using the principles of optimization inspired by the way living beings have evolved over three and a half billion years. The research goal is twofold: to understand better the adaptation processes of natural species and to conceive artificial systems using the same procedures. Evolutionary methods have proved very effective because they can resolve (in Holland's words) complex problems that "even their creators do not fully understand."[2]

It was after the 1990s, with the massive production of high-performance computers and the takeoff of the animat approach, that evolutionary methods were able to occupy a choice place among the adaptive methods applied to artificial systems. They were especially used in developing control architectures and (more recently) effective robotic morphologies.

The Best Way of Walking

In our laboratory, we enable the evolution of the neural networks monitoring the behavior of various robots, real and simulated. One approach that we follow is to code in these robots' genotype the rules of development of their nervous system. These rules are described in a program tree whose instructions indicate, for example, whether to create or suppress a neuron, to connect this neuron to another neuron, or else to change the parameters of its internal functioning. Some such programs are created by drawing at random the sequence of development instructions. These programs are going to generate the neural networks that will be the "brains"—control architectures—of the animats of the first generation. At each following generation, individuals that have the highest fitness and some others that have a lower one "reproduce" and transmit to their descendants new programs of development, in which the processes of crossover exchange one branch of the development tree of one of the parents for another branch coming from the other parent and in which mutations may change at random one instruction into another.

The implementation of this logic has enabled the production, in two successive stages, of neural controllers of locomotion and obstacle avoidance in Sect, our robot insect with six legs. The discovery of these behaviors was encouraged by an appropriate fitness and by the use of suitable sensors. Thus, in the course of a first stage in which the fitness of the robot was estimated by the distance covered in a given time, a neural network coordinating the movement of the legs was generated that allowed the robot to move in a straight line at the rhythm of the tripod walk.

The examination of the network produced by artificial evolution and generating this behavior reveals that it includes oscillators, or neurons that self-activate periodically in the absence of an external stimulus, such as the central pattern generators (CPGs) that biologists demonstrated in vertebrates' circuits of locomotion.

In the course of a second stage of development, another neural network enabled the robot to avoid obstacles encountered along its trajectory. This second network was able to connect with the first and to exploit the information provided by infrared sensors detecting the proximity of an obstacle. To obtain this result, we used a fitness that increased as the robot moved, but stuck at its current value when a collision with an obstacle did occur.

In simulation, an additional behavior—reaching a light source—was obtained by developing a third neural network that interfered with the two others. Such capacities were produced automatically in one night of calculations, whereas it had taken three thesis years for Randall Beer to conceive the neuronal controllers that allowed *Periplaneta computatrix* to exhibit the same behavior.

An Artificial Albatross

The Robur project[3] headed by Stéphane Doncieux at the AnimatLab—and later at the Institute of Intelligent Systems and of Robotics—also uses such evolutionary approaches. It has the goal of developing an autonomous flying animat with flapping wings—inspired by the albatross—of a sufficient size to carry the sensors, the electromechanic equipment, and the computing power necessary to ensure its autonomy.

In the current state of knowledge, nobody knows exactly what type of control to use for a robot with flapping wings or what ideal form to give the wings or what degrees of freedom to associate with them for this robot to fly effectively. This is why it is advantageous to let an evolutionary process adjust these characteristics.

To do so, an aerodynamic simulator permitting the modeling of a generic flapping-wing platform was conceived in order to be able to

Evolutionary Methods

There are several evolutionary methods, among which those grouped under
the term *genetic algorithms* are the most popular.

The general principle of these methods is to consider the problem as an
ecological niche in which individuals must survive and to consider these
individuals as more or less well-adapted solutions to this problem.

At each generation, many individuals propose solutions to the problem,
which some know a little better how to solve than others. The quality of each
of these solutions can be quantified by what is called its *fitness*. Like living
individuals, the proposed solutions have a genotype and a phenotype (1).
The genotype is generally made of a "chromosome" constituted of several
"genes", of which each represents a coded part of the solution, transmissible
from generation to generation (1*a*). The phenotype expresses this solution
and serves to evaluate its fitness (1*b*).

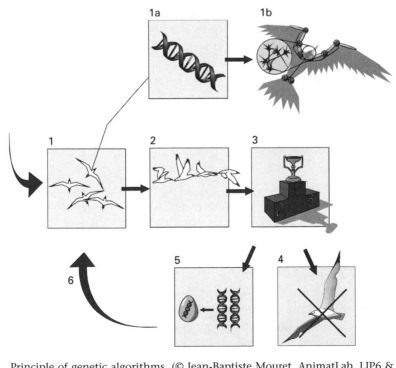

Principle of genetic algorithms. (© Jean-Baptiste Mouret, AnimatLab, LIP6 &
ISIR.)

The first generation of individuals is characterized by chromosomes with genes drawn at random (2). The coding of these genes is a delicate operation, for the efficacy of the algorithm strongly depends on it. Some individuals of this generation will be authorized to "reproduce." To select them, one draws at random the individuals having presented the best phenotype expressions—those having a good fitness—but also individuals that have presented less good ones (3, 4). In effect, the latter are necessary to maintain a genetic variability that enables the emergence of future good offspring.

During the reproduction of the selected individuals, their chromosomes and their genes can be modified due to genetic operators such as *crossover*, an operation of crossing by chance the chromosomes that come from two "parents,"[5] and *mutation*, which consists of randomly changing one gene into another (5). These two mechanisms enable the generation of solutions that are new compared to the parents' solutions and are likely to correspond still better to the problem being posed.

The chromosomes obtained by the parents' "reproduction" will provide the genotypes of the next generation, and the fitness of the proposed solutions will be estimated in turn. This process will be repeated until there is no longer any progress in the resolution of the problem under consideration (6). The individuals obtained in the final generation will be the better adapted, those that represent the best solutions to the problem posed.

Here we recognize (in a caricatured way) the principle of natural selection in Darwinian theory of evolution, but colored with the genetic discoveries of the twentieth century that the British naturalist was ignorant of.

compare various controllers and various morphologies. The bird's body was modeled with cones and cylinders. The wings were made of three rigid panels, with the internal and external ones able to move according to two degrees of freedom—the *twist* and the *dihedral* for the former, the *sweep* and the *twist* for the latter. The bird was to be equipped with motors to perform these movements and with a sensor to measure its speed. The evolutionary process was entrusted with discovering the structure of a neural network to coordinate the movements of the wing panels so as to ensure horizontal flight and constant speed, despite random perturbations of the surrounding air mass. Moreover, instead of optimizing in isolation from each other the energy consumed (the traction generated, the lift produced, and the engine's stability), evolution had to find a viable compromise among these different constraints. As observed in the course of natural evolution, the sought-for solution is a "good enough" solution—not the best possible for each of the criteria under consideration.

Figure 10.5
By evolution, Sect has learned to walk—as well as to avoid obstacles and to follow a light gradient. (© Jerôme Kodjabachian and David Filliat, AnimatLab, LIP6.)

This evolutionary process discovered two control strategies, one as effective as the other, both presenting similarities to the flight of real birds. In accordance with the first strategy, a homogeneous neural network is assigned to increase the amplitude of wing beats when the artificial bird needs to accelerate. In accordance with the second, the neuronal controller is divided into two subnetworks, one generating the beating, the other twisting the external panel to generate the necessary traction to maintain speed.

After having discovered the controls of the wing beats that ensures flight along a straight line, evolution was then left free to discover supplementary controllers acting on the two panels that constitute the bird's tail, enabling it to rise, descend, or turn to the right or left.

With the use of four additional sensors that inform the bird, first, about its angles of *roll* and *pitch* and about its *altitude*, and, second, about the direction of the objective to be reached, it was possible to produce automatically a neural network ensuring horizontal flight at constant speed toward this target. For this task, the second strategy of beating wings has proved more adapted than the first.

Although evolutionary methods managed to find how to beat wings while expending the least energy possible, it remains true that birds would

Vocabulary for a Flying Machine

A few definitions will enhance understanding of the explanations of the Robur project.

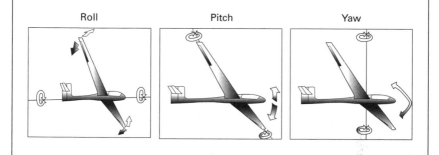

Roll   Pitch   Yaw

The roll, pitch, and yaw involve the whole machine. Roll is a rocking toward the right and the left; pitch is a rocking of the nose up and down; yaw is a rotation in the horizontal plane. All together, these three angles define a flying machine's *attitude*.

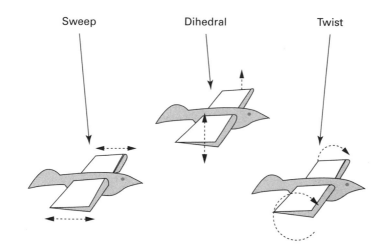

Sweep   Dihedral   Twist

In order to keep itself in flight, the machine can move its wings (or certain panels) back and forth (sweep) or up and down (dihedral), or it can twist them (twist).

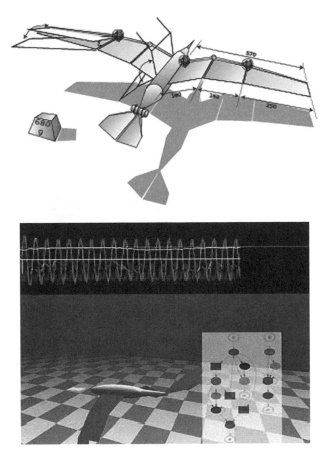

Figure 10.6
Flapping-wing flight. Top: The morphology of the virtual Robur. In particular, its wings are made of three panels, each with two degrees of freedom. By evolution, it is going to learn to control the movements of these panels to flap its wings effectively. Bottom: Robur's virtual environment and its evolved control architecture showing two subnetworks of neurons that control respectively the generator of wing beats and the twist of the external panel, which enable the animat to maintain constant speed and altitude of flight. (© Jean-Baptiste Mouret, AnimatLab, LIP6 & ISIR.)

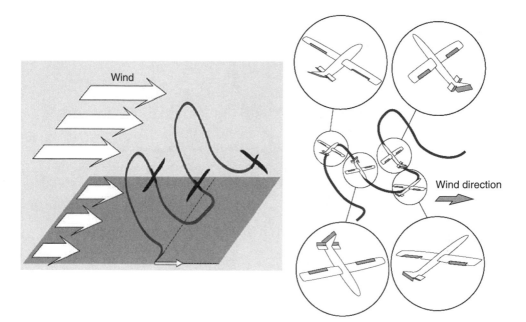

Figure 10.7

Soaring flight. Left: The albatross's dynamic soaring allows it to cover up to 6,000 kilometers in twelve days, practically without a wing beat. At the highest point of its trajectory, the albatross turns back from the wind and starts diving to gain speed. At the lowest point, the albatross sharply turns to face the wind and exploits the speed gained during the dive to gain altitude. Right: The trajectory of a simulated glider reproduces that of the albatross. (© Renaud Barate, AnimatLab, LIP6 & ISIR.)

be quickly exhausted by constantly using this mode of locomotion. So they have also learned to exploit the movement of air masses such as slope or thermal currents in order to alternate flapping and soaring flight over land, as do birds of prey. Other birds exploit the differences in wind speeds over the ocean surface—less above the wave surface than higher up—such as the albatross, which can glide along for whole days, expending almost as little energy as if it were resting.

Controllers capable of ensuring such behavior are automatically designed through artificial evolution in the framework of the Robur project. In particular, the albatross's characteristic trajectory that economizes energy by exploiting the wind gradient over the ocean is discovered by a simulated glider that seeks to exploit a minimum wind speed capable of maintaining it in flight over a given time period. This machine is equipped with three series of actuators—the *rudder* controlling the yaw, the *ailerons* controlling

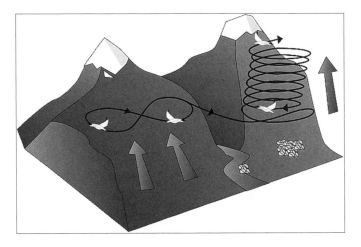

Figure 10.8
Soaring trajectories exploiting slope (*left*) or thermal currents (*right*). (© Stéphane Doncieux, AnimatLab, LIP6 & ISIR.)

the roll, and the tail *elevators* acting on the pitch—and with four sensors that measure its angles of roll and pitch, its altitude, and its heading with respect to the wind.

In the same way, the "eight-shaped" trajectories characteristic of terrestrial birds that exploit slope currents and the circular trajectories of birds that exploit thermal currents have also been discovered through the artificial evolution of simulated gliders.

Swim and Walk Like a Salamander
It is not easy to refine even a single mode of locomotion for a robot. Robur does not yet know how to alternate between flapping and soaring flight in an adaptive way. Likewise, MLR III, the Japanese gibbon-gorilla robot, cannot yet autonomously select which one among its three possible locomotion modes—brachiation, quadruped walking, and biped walking—is the best adapted to current circumstances.

A roboticist from the Bioinspired Robotics group at the EPFL and a neurobiologist from the Institut national de la santé et de la recherche médicale in Bordeaux (Auke-Jan Ijspeert and Jean-Marie Cabelguen) have achieved this kind of behavioral switch. They managed to obtain a single controller that triggers both an anguilliform swim when their robot is in the water and a quadruped walk when it is on dry land. The control architecture of this robot, called *Salamandra robotica*, is inspired by the salaman-

Figure 10.9
Salamandra robotica immersing itself in Lake Leman. (Photograph by A. Herzog, courtesy Biologically Inspired Robotics Group, École polytechnique fédérale de Lausanne.)

der's nervous system. This animal is the ideal model for this study, for it swims like a primitive fish and moves on land like a lizard.

Neurobiologists have demonstrated that in the real salamander, the two groups of CPG neurons situated in its spinal cord and commanding, respectively, the members actuating walking and the undulatory swimming movements of the body are not entirely independent, as had been previously thought. Of course, stimulating the CPGs of one or the other produces movements of walking or swimming. But by simulating them with increasing intensity, the first group of neurons becomes saturated, and the legs become immobile, whereupon the second group takes control, and the salamander's body starts to undulate.

Salamandra robotica is composed of modules linked by flexible joints and equipped with four rotating legs. The swans of Lake Leman have already been able to observe it going into the water and reappearing on the banks. Its control architecture is closely inspired by the circuits linking CPGs to the motor neurons of the salamander's spinal cord, and it has in effect evolved by taking neurobiological knowledge into account. This work has verified that by increasing gradually the intensity of the electric signal in the CPGs, the robot moves easily from walking to swimming— and vice versa. This same mode of stimulation also permits a modification of the speed and direction of locomotion.

This study illuminates the way the nervous circuits in amphibians have been modified by evolution to allow them to leave the water and adapt to dry land. According to the researchers, the model might be applied to all tetrapod vertebrates, including humans. Therefore, it might help doctors and neurologists to understand certain patients' motor problems.

Coevolution

The Paradox of the Red Queen
Alice: "Well, in *our* country, you'd generally get to somewhere else—if you ran very fast for a long time, as we've been doing."

The Red Queen: "Now, *here*, you see, it takes all the running *you* can do, to keep in the same place."

In *Through the Looking Glass*, Alice and the Red Queen are moving at the same speed as the countryside around them. This would be the consequence of a coevolutionary process, forcing living systems into an endless arms race. When some systems invent new equipment or behaviors to increase their resources at the expense of other systems, the latter invent equipment or behaviors to counter them, and so forth. The paradox resides in the fact that the complexity of these inventions seems to increase in the course of evolution, but without always corresponding to an increase in the quality of their owners' adaptation to the environment.

The emergence of very diversified behaviors might be explained by living systems' incessant confrontation with a dynamic environment populated with creatures pursuing the same survival objectives. Numerous simulations of such coevolutionary process have been undertaken.

Researchers at the EPFL Intelligent Systems laboratory have studied the coevolution of predator and prey behaviors among robots—in this case, small wheeled robots. The predator has a visual sensor to detect prey at a distance, and the prey is blind but can move twice as fast as the predator.

Both have infrared sensors to detect at short range a wall or another robot. The predator's fitness is based on its capacity to catch the prey, and the prey's on its capacity to escape the predator.

Two populations of one hundred individuals have involved over one hundred generations and have demonstrated that coevolution leads to limitations in the two groups' performances—a reminder of the Red Queen! In effect, confronted with adversaries prevented from evolving, the predator would manage much more effectively to trap the prey, and the prey would manage much more effectively to avoid predators. However, coevolution entailed the emergence over generations of several different strategies: the predator first trapped the prey by pursuing it, then changed tactic and lay in wait until pouncing on it when it approached; similarly, prey were at first immobile near a wall that the predator avoided and then fled as soon as the predator arrived, finally choosing to move rapidly and sometimes randomly so that their trajectories were not too predictable. Each of these strategies was abruptly discovered when the adversary had sufficiently evolved to render the previously adopted strategy no longer effective. Some strategies could reappear in the course of evolution.

The conclusion of this work demonstrates that, overall, the individuals that resist the best over time are those that have developed simple and easily modifiable strategies rather than those that have developed complex strategies that are able to serve in many different circumstances. To be able to choose rapidly from a repertoire of simple strategies would be far more adaptive for animats than to elaborate a single very sophisticated strategy. This finding confirms some conclusions formulated with regard to swarm intelligence.

Coevolve to Communicate

This same team, working with Laurent Keller, a biologist at Lausanne, has also investigated some factors influencing the evolution of communication in societies of insects. These researchers first caused in a simulation the evolution (under different selective pressures) of ten colonies of robots over five hundred generations; then they downloaded the corresponding architectures to monitor the actual robots, "s-bots," having the same characteristics as the simulated robots.

The robots were placed in an environment where sources of "edible" or "poisoned" food were placed (in actuality, zones surrounded with either pink or red lights). The s-bots were equipped with visual sensors capable of discriminating between the two types of resources. They could emit a blue light and perceive the blue light emitted by another robot. Their

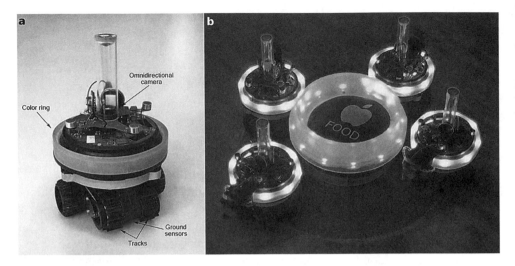

Figure 10.10
S-bots discovering a source of edible food. (© Dario Floreano, École polytechnique
fédérale de Lausanne.)

control architecture possessed only thirty or so neurons whose parameters
were at first drawn at random. Evolutionary processes were going to adjust
the ways in which the robots were moving, in which they emitted a light,
and in which they were going to "interpret" the lights emitted by their
fellows so that they survived as long as possible. Communication between
robots—which was not programmed in advance—could be translated into
an emission of blue light to signal a resource. Robots could then interpret
the signal in terms of either approach, if the resource was edible, or avoidance, if it was poisoned. The problem was that access to an edible resource
was limited, for the robot had to be in direct contact with this zone to
benefit, and thus a robot had no "interest" in signaling this resource to
others.

Two kinds of colonies were tested: one made up of robots having
kinship relationships and hence close genotypes ("related"), and the other
made up of robots with different genotypes ("unrelated"). Two modes of
selection were implemented, one keeping in each generation the *individuals* that had the best survival time in each of the colonies (individual criterion) and the other keeping in each generation the robots of *colonies*
surviving the best (group criterion). All the robots were initialized with
totally random behavior; those that most avoided the sources of poison
were allowed to reproduce. In analyzing the results after five hundred

generations, the researchers realized with astonishment that despite the simplicity of the s-bots' control architecture, they had developed differentiated communications. In fact, when the robots were related and the selection was performed at the group level, communication developed rapidly and was used with a view to cooperation: certain individuals specialized in the emission of signals announcing the proximity of food, and others in signaling poison. The survival of these colonies was greatly improved. In contrast, when the robots were not related or when the selection was performed according to the individual criterion, communication was very poorly developed. When it existed, it could lead to a drop in overall performances: certain individuals might even develop deceptive signals, leading some of their fellows far from sources of food in order to consume more themselves and to prevent hungry individuals from reproducing. This might be a very primitive form of what some call "Machiavellian intelligence."[4]

Compared with the learning experiences of the Talking Heads, this work brings additional information about the influence of the "social" structure and the mode of selection (whether individual or group) on communication—which communication is certainly very primary in this colony of robots. Laurent Keller, whose specialty is the study of ants, sees here strong analogies with what can be observed among social insects. In effect, cooperative communication that has favored the transmission of similar genotypes among related robots reminds him of how the worker ants *Formica exsecta* maintain "normal" relations with the ants that are the closest to them genetically but eliminate those ants that are less close.

Coevolve to Learn

The collective experiment with the small robots called Cyber Rodents at the Science and Technology Institute in Okinawa (under Kenji Doya) is rather similar. These robots possess omnidirectional vision and infrared sensors to detect obstacles and fellow robots, and they can emit three types of light for possible communication. Their overall objectives are to "survive," , i.e., be able to recharge on batteries scattered in the environment, and to "reproduce," i.e., be able to exchange their chromosomes with other robots via an infrared port. The batteries have different colors and represent several types of resources, attractive and repulsive, but the robots do not know the meaning of these differences at the beginning.

The interest of this work lies in the fact that it is not precise behaviors that are supposed to emerge in the course of generations, but rather *procedures for learning* such behaviors. Thus, the evolutionary process adjusts

Figure 10.11
Cyber Rodents survive and reproduce in a laboratory environment in which several energy resources are placed. (© Kenji Doya, Okinawa Institute of Science and Technology.)

the learning of the nature of the reward, the use of one or several modes of learning, the updating speed of the memory, the time spent in random search for new elements, and so on. This work demonstrates that artificial evolution can aid a roboticist not only in conceiving the structure of a control architecture, but in determining the way in which the structure will be improved by learning.

Coevolve *and* Learn
The European project Swarm-bot, coordinated by Marco Dorigo, marks a breakthrough in the increasing complexity of tasks to be performed. Thirty-some small s-bots were provided with grippers that allowed them to hook onto each other and small lights to communicate with the closest robots. By alternating evolutionary and learning methods, the s-bots proved

Figure 10.12
Several s-bots attached to each other manage collectively to cross a gap, which an isolated robot cannot do. (© Michael Bonani and Francesco Mondada, École polytechnique fédérale de Lausanne, Project Swarm-bots.)

capable of autonomously aggregating themselves in different shapes depending on the task at hand: pushing a heavy object while avoiding obstacles, gathering objects into an arena, or even crossing a gap—different tasks that no robot in isolation could accomplish. Setting aside the spectacular swarm robotics results obtained, this study had the main goal of testing the influence of group size on the various tasks to be performed.

Coevolution of Morphology *and* Control

Evolutionary methods have proved successful in elaborating a good control architecture for a given robotic body. The roboticists' hope is to be able to submit simultaneously to the evolutionary process the design of both the morphology and the associated control architecture.

Karl Sims, an American researcher, obtained spectacular results a few years ago by evolving both the shapes and the controllers of improbable creatures. Constituted by the random combination of parallelepipeds of different sizes, relatively variable in number and in arrangement, these creatures were placed in an aquatic or land environment in which they proved capable—sometimes in a comical way—of swimming or walking or

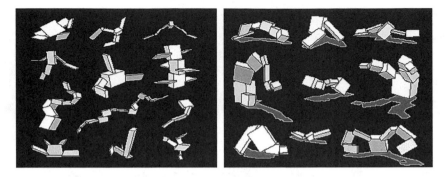

Figure 10.13
Some improbable creatures that have evolved by adapting to swimming (*left*) and to walking (*right*). (© Dunod).

jumping. Naturally, just as an elephant and an ant do not control their limbs in the same way to walk, a particular controller was associated by evolution with a particular morphology.

In the framework of the Genetically Organized Lifelike Electro Mechanics (GOLEM) project at Brandeis University under Hod Lipson and Jordan Pollack, an additional evolutionary stage was reached by going from simulation to reality. Taking advantage of a fitness criterion that was defined beforehand, the corresponding creatures could evolve not only in their morphology, but also in the structure and parameters of their control architectures. In order to produce machines capable of advancing on the ground, a process of artificial selection was conducted in simulation, as Sims had previously done. At the end of this process, the genotype of the most successful individual was supplied to a rapid prototyping machine, a sort of three-dimensional photocopier, which decoded the corresponding data and produced, layer by layer, the phenotype previously selected by artificial evolution.

The researchers began by tackling this robotic autoconstruction in the simplest possible way. For animating these creatures, they put at the creatures' disposal building blocks in the form of bars of different sizes as well as joints and actuators acting on these joints that would likely connect and animate the building blocks. The corresponding fitness assessed a creature's capacity to move in a straight line on firm ground. Based on a population of two hundred individuals, six hundred generations were sufficient to produce individuals that moved correctly.

Despite the simplicity of the problem and of the morphologies likely to solve it, the authors were surprised to note that many solutions were dis-

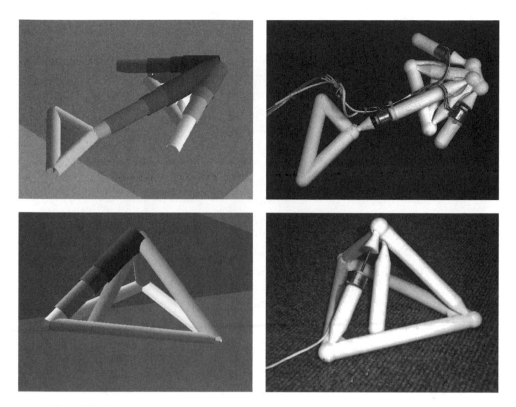

Figure 10.14
Left top and bottom: Simulated forms conceived by artificial evolution. Right top
and bottom: Real robots produced according to the plans of the virtual robots. (©
Jordan Pollack, Brandeis University.)

covered, all with a similar efficacy. Symmetric structures often emerged
although no instruction had imposed such findings from the start. Other
solutions proved robustly resilient to changes in the length of the compo-
nent bars, changes that did not affect their mobility.

These results, as interesting as they may be, should not obscure the difficulties of the enterprise. On the one hand, the conception of a robot that
has evolved in a virtual universe is not always perfectly adapted to the real
world simply because one does not know how to model all the aspects of
this world. On the other hand, it is not certain that the concepts, algorithms, and machines that are implemented in this research are adapted
to the resolution of problems that are more complex than those that have
been tackled up to now. In particular, to overcome the current limitations,
the Brandeis researchers had to launch a vast mobilization campaign,

asking individuals to host calculations in parallel on their personal computers; thirty thousand volunteers came forward. But the researchers had to put an end to this operation, explaining to their volunteer partners that the methods used were not propitious for enlarging the field of possibilities. They are now orienting their investigations to come up with new theories and new evolutionary mechanisms in the hope of breaking through the "complexity gap" in which they have been trapped.

Let us mention another study undertaken by the same team that opens up important prospects for robots because it concerns their behavioral reorganization in case of breakdown or "injury." One of their robots, a sort of starfish with four branches, managed to reorganize its movements when it lost one of its arms. To do so, it began by forging through experiment an "internal model of itself," for it did not "know" what form it had. By relating its sensation to different possible forms, it elaborated several models of itself and retained the one that appeared to it most probable. It could then exploit this model to generate new behaviors. When one of its constitutive elements no longer functioned, it "mentally" forged other models on the basis of the model that had worked well up to then, and it used the one that best corresponded with its new morphology to try to accomplish its mission regardless.

Thus, research aiming at increasing the autonomy of artificial systems by drawing on biomimetic inspiration is well advanced in various laboratories in the world. To catch up after a delay in this domain, the European Union has recently launched a new research program, "Future and Emergent Technologies," to last until the year 2020. It particularly encourages work involving robots capable of self-configuring, self-optimizing, self-protecting, and self-healing.

III   Hybrids

11   From Prostheses to Cyberprostheses

There are more things in heaven and on earth, Horatio,
than are dreamt of in your philosophy.
—William Shakespeare (1600)

Antiques

Ancient traces prove that humans in all ages sought to replace organs or limbs that were deficient or lacking.

In the Rig-Veda, the sacred texts of Brahmanism dating from 3,800 to 2,700 BCE, it is written that during a battle, the ferocious warrior Vispala was wounded and that her leg was amputated "as severely as the stump of a wild bird's wing." The Asvins, heavenly twins and medical experts, gave her an artificial iron leg so that she could take part in the next battle. Similarly, 2,400 years ago, Herodotus recounted that the Elean soothsayer Hegesistratus, having been put in irons by the Spartans, grabbed a sharp object lying around and cut off "part of his foot before the toes, after having found he might pull the shackles off the rest of his foot." After escaping by making a hole in the wall, he made himself a wooden foot. History does not say how he was miraculously cured of his amputation.

Prostheses of this type dating from around 2,200 years ago have actually been discovered, such as an articulated hand on the arm of an Egyptian mummy, an artificial foot in Kazakhstan, and dental bridges and implants throughout the Mediterranean world. But the artificial eyes so superbly painted or sculpted by the ancient Egyptians were probably used only to decorate statues.

Passive Prostheses

Other more recent stories of prostheses traverse the centuries. In the six-teenth century, during a stay in Germany, the Danish astronomer Tycho Brahe at the age of twenty quarreled with a compatriot over Pythagoras, which ended in a duel in which part of his nose was torn off. He was fitted with a nasal appendix made of gold and silver—or perhaps simply of wax.

Other prostheses, a little less artisanal than this one, were described for that era. The German knight Goetz von Berlichingen lost a hand at the siege of Landshut in 1504 and had a blacksmith near his chateau in Baden-Wurttemberg make him one with four articulated fingers, an apparatus that still exists, conserved in the chateau museum. A few years later the Swiss watchmaker Charles Cusin is said to have made an articulated arti-ficial arm with a hand that could open and close.

The master here was no doubt the famous French surgeon Ambroise Paré—helped, we daresay, by the sixteenth-century introduction of new means of warfare, such as mobile cannons and light artillery, which caused a great number of injuries. By inventing the ligature of arteries, he paved the way to healing amputated limbs.

Figure 11.1
Left: Tycho Brahe and his artificial nose. Right: Goetz von Berlichingen and his prosthetic hand. (© Dunod.)

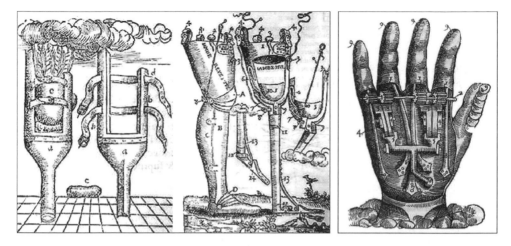

Figure 11.2
Prostheses made by Ambroise Paré. Left: The "poor man's leg" and the "rich man's leg." Right: An articulated hand.

Ambroise Paré specialized in articulated prostheses and built both a wooden arm that pivoted at the elbow thanks to a spring and then a hand with mobile fingers. He also conceived two models for legs, one a "poor man's leg"—a rigid pylon—and the other a "rich man's leg"—articulated at the knee and heel and with a more natural shape. We also note that it was Ambroise Paré who devised the first ocular prostheses, heavy balls of enameled gold and silver. They would be replaced only a century later by lighter prostheses made of glass, whose secret of fabrication was long jealously guarded by the Venetians.

Alongside the construction of bioinspired automata in the seventeenth century, mechanical prostheses were improved, aiming at the quality of function more than at the aesthetics of form. The principle of these prostheses is still used today, though modernized by the nature of new materials.

Era of Cyberprostheses

The end of the twentieth century saw other surgical prostheses appear: the replacement of internal organs by *endoprostheses* in the body, such as cardiac stimulators, vascular stents to help dilate an artery, and artificial cartilage and rigid structures for knees and hips.

Recent progress in electronics enables the conception of "intelligent" prostheses and endoprostheses, or *cyberprostheses*. These new prostheses are distinguished from those previously described ("passive" prostheses) because they are controlled by a microprocessor linked directly to the bearer's nervous or muscular system. The endoprostheses are electronic chips introduced into the patient's body, but without being linked to an artificial organ. They are used either to send motor orders to a natural organ or member or to process information received by natural sensors.

However, these interactions at contact between the living and the artificial do engender other problems connected to their reciprocal compatibility. The engineering of biomaterials (defined as "nonliving materials used in a medical device intended to interact with biological systems"[1]), which includes metals and metal alloys, ceramics, and synthetic or natural polymers, is in full expansion. Some of the inventions cited at the beginning of this book, such as those inspired by the spider's web and the blue mussel's glue, are designed to improve these interfaces between living and artificial.

Later on we give some examples of cyberprostheses—artificial systems integrated into the living—even if they are still far from turning those who wear them into the bionic heroes of pop fiction. Such hybridizations involve not just living systems: for the first time in the history of human inventions, except for the already mentioned example of the automaton chess player made by von Kempelen, some labs have tried the inverse—that is to say, integrating the living into the artificial. They are trying to come up with kinds of bionic robots, a few examples of which are described in the following chapter.

12   Hybridizing the Artificial

If we take nature as a guide, we will never go astray.
—Marcus Tullius Cicero (44 BCE)

After three and a half billion years of evolution, some of the materials, structures, and procedures that nature has invented and perfected have proved impossible to copy. In effect, we have not yet understood how to reproduce their properties—or else we do not yet have the tools to do so. So the temptation is great to recuperate directly the part or the organ of a given living system that is responsible for the properties in question and to incorporate this part or organ into an artificial ensemble that would then benefit from these properties. Naturally, one must maintain this "implant" in a functioning state within the hybrid system, which, although not simple, is not always out of the question. This logic is already being exploited in several different ways, and it appears likely that it will spread because both the fundamental and applied fallout of such efforts appear so promising.

Hybridization with Natural Sensors

Originating in China, the silk moth was domesticated around 2,600 BCE. Its larvae stuff themselves on mulberry leaves, and in the space of thirty to forty days they become twenty-five times more voluminous and twelve thousand times heavier than at birth. At the end of their growth, they cease feeding and activate two saliva glands situated behind their mandibles that secrete an uninterrupted filament of silk that composes the cocoon housing the nymph. The butterfly that emerges from the cocoon is an insect of heavy flight, hindered by its weight and the small size of its wings. It lives only a few days, in the course of which it is occupied with ensuring the survival of its offspring. To do so, the virgin female emits a pheromone called *bombycol*, a single molecule of which suffices to excite the olfactory sensors of the male silk moth antennae and to trigger its behavior to locate the female. It has been shown that the male reacts to a concentration of bombycol as weak as two hundred molecules per cubic meter of air and that it can find a female up to 10 kilometers away!

The construction of an odor detector of such sensitivity will long remain beyond the scope of human technology. This is why Japanese researchers have taken the antennae from a male silk moth and maintained their functionality by extending their extremities with two glass capillaries filled with a solution of physiological serum to temporarily prevent them from drying out. The researchers then connect these antennae to an electronic apparatus via copper wires crossing the capillaries and then place them on the front of a little wheeled robot that is almost as small as the insect.

Thus, the nervous influx emitted by the antennae have been translated into "electroantennograms" that make it possible to monitor the voltage difference between the extremities of each antenna. When this difference exceeds a certain threshold, the presence of bombycol is detected, and this information is transmitted by a computer to the motors governing the robot's wheels. A very simple program then enables the robot to exhibit a chemotaxic behavior and to pursue the panache of the pheromone that the experimenters vaporize in front of it. This program is made up of the following instructions:

• If the right/left antenna detects bombycol, then make the left or right wheel turn so that the robot turns toward the right or left;
• If the two antennae detect bombycol, then turn the two wheels so that the robot moves straight ahead;
• If neither antenna detects bombycol, stop the wheels so the robot does not move ahead.

The similarity of behavior exhibited by the robot and by the insect allows a better understanding of why surprising zigzags characterize the insect's trajectory.

If current technology does not allow us to keep the silk moth antennae in a good working state beyond one day or able to detect the source of the odor beyond a few centimeters, it remains true that the future of this type of research appears promising with respect to all the natural sensory arrangements that have defied this technology, but yet worrying because of the ethical problems that it raises.

Figure 12.1
Left: A male silk moth's antennae. (© Dunod.) Right: A robot equipped with a silk moth's two antennae (arrow). (© Ryohei Kanzaki, University of Tokyo.)

Hybridization with Natural Actuators

Another challenge that current technology is far from meeting is to produce actuators that present the qualities of plasticity and "reparability" of living muscle tissue. The energy yield of this tissue has proved excellent because it can produce 1,000 joules of work per gram of glucose consumed. Moreover, it is consuming a renewable resource and is producing waste that is absorbable by the environment.

With a view to evaluating the difficulty of using such a material as a robotic actuator, the U.S. federal Defense Advanced Research Projects Agency (DARPA) has subsidized the design of an aquatic robot equipped with a tail made of synthetic elastomere that is able to be activated by a pair of muscles taken from a frog. The movements of the tail are produced by the antagonistic contractions of the two muscles, themselves provoked by electric impulses controlled by a microprocessor. To maintain these natural implants in a functioning state, the robot is bathed in a solution of physiological serum that includes antibiotic and antimycosic agents as well as the glucose that supplies the muscles with the necessary energy. It has proved possible, over a survival period of forty-two hours, to make the robot perform a total of four hours of simple maneuvers: starting up, stopping, turning, and swimming in a straight line, at a speed reaching one-third of its length per second.

Here again the first results are encouraging, though several problems remain to be resolved before real practical applications can be hoped for. In particular, researchers will have to find a means of maintaining the muscular implants for longer in working order—and more often. It will

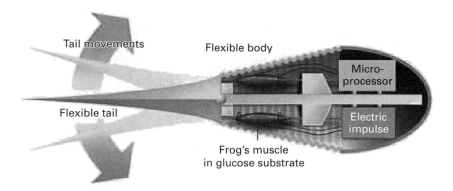

Figure 12.2
A robot fish controlled by frog muscles. (© Dunod.)

also be necessary to be able to modulate the excitation of a greater or lesser number of muscle fibers depending on the movement to perform, rather than submit the whole muscle to the same electric field, as was done here.

Hybridization with Natural Control Architectures

Nature has invented many ways for living beings to react to the modification of their internal and external milieus. Single-cell systems, for example, use sensors and actuators at the molecular level that communicate by chemical or mechanical means. More evolved animals use differentiated organs and specialized tissues (such as the nervous system) to send signals of an electrical nature. Each use has its own logic and adaptive properties— both of which various research projects have tried to import into hybrid systems.

A Single Cell

Physarum polycephalum is an organism difficult to classify, lying between an amoeba and a mushroom. This sort of yellow mildew is formed of several cells fused into a single, amorphous organism, its size varying from a few tens of microns to several meters. Its exchanges with the environment take place through an endoskeleton made of a very dense network of microscopic tubules filled with cytoplasm. It is by the transport of substances that organize themselves in this network, as a function of the stimulations received, that *Physarum* can perceive and also act: it can thus capture the odor of bacteria and approach in order to absorb them, and it can capture a source of light or move away from it. It is the latter property that is being exploited to control a robot.

A team of Japanese and British researchers have succeeded in seeding with a few extracts of *Physarum* six zones of a physical base covered with agar, a gelatinous substance often used as a culture milieu in microbiology, then in making the zones grow so that they reconstitute a single star-shaped cell. When a light source projects a beam onto one of the branch extremities, this stimulation has the effect of making the cytoplasm contained in the tubules of that area oscillate. A six-legged robot is equipped with six light detectors, each of which is in contact with one of the branch extremities of the organic circuit. According to which sensor is stimulated, one of the robot's legs is activated by the oscillation of cytoplasm. In other words, when the robot's sensors detect a light source in a certain direction, the *Physarum*'s oscillations are translated into a motor pattern that makes the legs move so that the robot withdraws from the light.

The advantages of such an assembly are numerous and will probably generate other such research and applications. In the first place, just as the corresponding control structure is capable of self-organizing on the basis of six sources of initial seeding, so this structure will be capable of repairing itself if, for example, one of the organism's arms is cut off or destroyed. Similarly, the energy cost of functioning is particularly low, and the system can function for several days without being fed. Finally, it can be maintained dry for a long time in a dormant state and be reactivated by humidification.

A Group of Neurons in Culture

At Florida University, researchers under the direction of Thomas DeMarse have made a solution containing about twenty-five thousand neurons obtained by the mechanical and chemical destructuring of a portion of the cortex taken from a rat embryo. They then poured this solution over a substratum containing a grid of sixty-four electrodes covered with a membrane enabling the observation of the preparation while avoiding the risk of bacterial infection. At the end of three to five days, the neurons within this culture spontaneously established connections among themselves. After ten days, the neural network thereby developed could generate a synchronized neuronal activation characterized by semiperiodic bursts of electrical activity. These bursts were then recorded and analyzed by means of the substratum's electrodes and persisted the whole duration of the network's life.

Figure 12.3

A robot controlled by a mold. Upper left: A portion of *Physarum polycephalum* with the network of tubules that traverses it. Upper right: *Physarum* in a star-shaped configuration. Lower left: The apparatus that controls a hexapod robot by means of a *Physarum* cell. Lower right: The hexapod robot. (© Soichiro Tsuda, University of the West of England.) See color plate 19.

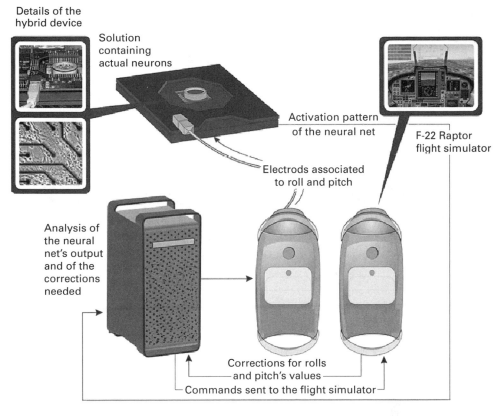

Figure 12.4
The control of an F22 Raptor by means of a neural network in culture. (© Thomas
DeMarse, University of Florida.)

This rudimentary neural network is then connected to an artificial
system that is particularly complex: the flight simulator of an F-22 Raptor!
The goal is for this network to control the engine's roll and pitch while
maintaining it in horizontal flight. To do so, two of the sixty-four electrodes are used not only to measure the average electrical activity in the
part of the network that surrounds each of these electrodes, but also to
send electrical impulses to the two corresponding zones of the network.
Under these conditions, at regular intervals an electronic montage enables
the network to be stimulated at the level of each of the electrodes, according to the respective amplitude of the correction need for the angles of roll
and pitch. The pattern obtained after a stimulation of the electrode associated with roll modifies the angle of the ailerons; the pattern associated

with the pitch acts on the rudder. These stimulations have the effect of modifying the network's synaptic connectivity, allowing a learning process. In effect, as long as the plane's attitude needs to be corrected, the corresponding impulses modify the neural controller's synaptic weights and consequently the motor orders it engenders. But once this attitude is stabilized, the controller is no longer stimulated and does not change any of the commands.

In practice, the results obtained are encouraging, but some instabilities still need to be corrected. Equivalent approaches have been used to control real robots, whether robotic arms that produce artistic drawings or wheeled robots that avoid obstacles or follow each other. The control of a glider's attitude on the basis of images gathered by an onboard camera has even been tried, with results as yet unknown.

Nevertheless, as of today, the behaviors that are controlled are simple. It seems probable that such approaches will lead to greater difficulties when the control of more complex behaviors is sought in the future.

An Encephalon in Culture

Beyond an isolated cell or several thousands of neurons spontaneously reorganized into a network, a whole functional part has been taken from an animal's nervous system and hybridized in a series of experiments conducted at the University of Chicago.

The part of the brainstem that controls the lamprey's attitude, which transforms data coming from the vestibular system and other sensors into motor orders that stabilize the body's orientation during swimming, has been used to control a wheeled robot's reaction to light. In the corresponding setting, the signals captured by the robot's light sensors have replaced the data on attitude used by the animal, and the motor orders serving its equilibrium are instead being used to control the rotation of the robot's wheels.

According to the state of the electrodes that either excite the nervous tissue or else record its electrical activity, it was possible to implement in the robot a reflex of either pursuit or else avoidance of light in order to study the adaptive capacities of this tissue. It was observed that when the robot was confronted for a few minutes with an asymmetric luminous environment (more light coming from the right than from the left, for example), some long-term adaptations occurred in the controller's synaptic connectivity. In the case of a setting implementing a pursuit reflex, the left wheel turned faster at the start than the right wheel, so the robot moved toward the more luminous zone. But after the robot became used

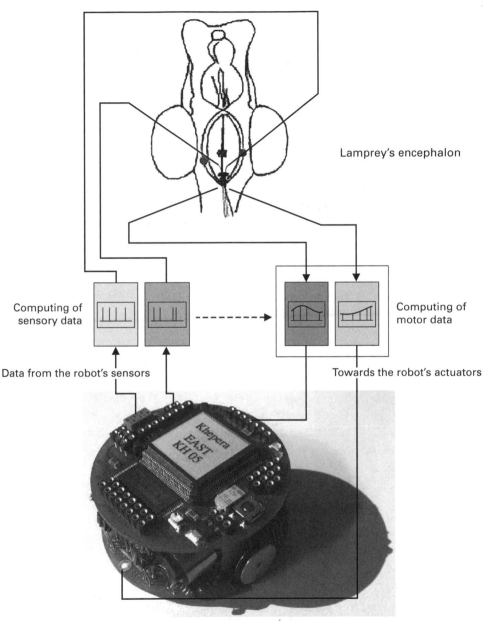

Lamprey's encephalon

Computing of
sensory data

Computing of
motor data

Data from the robot's sensors

Towards the robot's actuators

The Khepera robot

Figure 12.5
An encephalon from a lamprey (top) controls a small mobile robot's reactions to
light (bottom). (© Dunod.)

to the asymmetry of the light environment, the difference in the wheels' rotation speeds diminished: the robot was habituated to this situation.

Like traditional neurophysiological experiments, this approach, according to the researchers who practice it, offers new, useful, and complementary information about the way the nervous system adapts and learns.

Other Hybridizations

At the Microbe Level

The autonomy of a living system is of course linked to the efficacy of its sensors, its effectors, and the way its control architecture connects the two to each other. However, the capacity to produce the necessary energy for its proper functioning is just as indispensable in order to ensure its energy autonomy.

To perform missions of long duration in an unknown environment—unprepared by humans—the robots of the future will not be able to capitalize on the availability of devices for recharging their batteries. A technology to manufacture a system of electricity production that is inexhaustible, autonomous, and sufficiently compact to be loaded on board a robot unfortunately does not yet exist—hence, the idea of searching in the living world, among bacteria, for a way to resolve this problem.

At the University of Southern Florida, for example, batteries of microbial fuel, exploiting the good offices of the bacteria *Escherichia coli* (such as those that populate our own stomachs) have served to produce electricity to fuel Gastrobot also nicknamed Chew-Chew, a sort of train with three wagons. The first one houses a fuel battery filled with a colony of bacteria. By a tube situated above the battery, morsels of sugar are little by little sent to the artificial stomach. There, by decomposing the carbon chains, the bacteria liberate electrons to recharge (through oxidoreduction) the batteries situated in another wagon. This clever arrangement is not yet operational: to recharge its batteries totally, the robot needs three sugars, which it takes no less than eighteen hours to digest!

Related to this "sugarvore" robot, a carnivorous engine has been developed in Bristol by Chris Melhuish's team. Slugbot was made with an articulated arm at the end of which is a projector, a gripper, and a camera, thanks to which it can find and capture slugs that are found in its environment—up to a hundred per hour. The objective was for this robot to produce its own electricity by digesting these slugs in a "stomach" similar to that of Gastrobot. This type of prey was chosen because it is highly energetic and relatively easy to identify (it glows in the light), and it cannot

Figure 12.6
Robots that "digest." Top: Chew-Chew, the sugar-eating robot, and its inventor. (©
Stuart Wilkinson, University of South Florida.) Bottom left: Slugbot, the slug hunter.
Bottom right: Ecobot II, the robot that digests flies. (© Chris Melhuish, University
of Bristol and University of the West of England.)

rush off at the approach of a predator. In certain demonstrations, in order
not to hurt sensitive souls, the slugs were replaced by bits of ripe banana!
Whatever the case, the financing for these experiments ran out before the
slightest slug could produce the slightest electron.

Although these first approaches were remarkable, they nevertheless were
abandoned because their energy yield appeared too low. This is why the
future of research on robots' energy autonomy now rests on an improved
technology that capitalizes on the use of an oxygen cathode to extract the
electrons. Thus, Ecobot II, another Bristol robot, is equipped with a series
of batteries composed of microbial fuel housing bacteria taken from the

mud of a nearby purification station. These batteries generate electricity from the flies fed to the robot. The battery bacteria decompose the glucides contained in the insects' exoskeleton, which produces sufficient electricity to enable the robot not only to move around, but to do useful work such as directing itself toward light and transmitting by radio to a remote station information on the exterior temperature.

The corresponding performances are as yet quite modest: the robot moves only an average of two to three seconds every fifteen minutes and over a distance of 2 to 3 centimeters before being obliged to stop to accumulate the energy necessary to continue moving, measuring, and sending information. Nevertheless, it can work for five days continuously by digesting a single fly in each of its energy batteries.

The work in progress leads to hope for an improvement in the energy yield so that the robot can itself seek and trap the flies it needs by using odor traps or appropriate lights. Other sources of energy rich in sugar, such as fruits or the shells of crustaceans, might also be used.

At the Molecular Level

In 1959, Richard Feynman, a future physics Nobel winner (1965), announced that it would one day be possible to construct a machine atom by atom with the help of other machines. In 1986, less than thirty years later—the same year that the Scanning Tunneling Microscope that permitted the "sight" of individual atoms was perfected[1]—Eric Drexler published *Engines of Creation*. This work founded at the intersection of atomic physics, engineering, and science fiction a new field of research called *nanotechnology*, which we mentioned in part I for its contributions. The book dramatizes an apocalyptic scenario in which nanomachines escape any control and swallow the world. Although as such it continues to feed scientific controversy, political argument, and public debate, it nevertheless has been a source of inspiration for more peaceful and quite real nanomachines that try to integrate biomolecules with man-made nanostructures.

This is the case with "the smallest helicopter in the world," conceived by a team from Cornell directed by Carlo Montemagno, thanks to the use of the F1-ATPase, an enzyme particularly widespread in the living world and whose reversible motor properties have been known since 1967. Including a rotor in its central part, this assemblage of proteins only 12 nanometers in diameter is encrusted with a lipid membrane of mitochondria that participates in the synthesis of adenosine triphosphate (ATP), the molecule that gives living systems the energy they need. One barely understands the mechanism that turn this "motor" at a speed of about 130

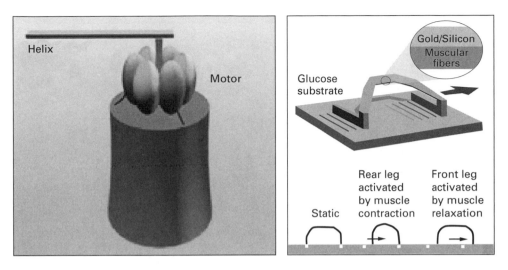

Figure 12.7
Nanomachines. Left: A nanohelicopter with a "motor" of F1-ATPase. Right: Nano-robot climbing, thanks to a rat's muscular cell. (© Dunod.)

rotations per second: it seems that it involves a gradient of protons generated on each side of the membrane by specialized pumps. Each time this mechanism forces the ATPase rotor to turn 120 degrees, an ATP molecule is synthesized in the cell. Inversely, the rotor will start to turn spontaneously if the molecule is fed with ATP, and it is this property that American researchers have used to create a nanohelicopter by giving the rotor a tiny propeller made of nickel!

This same team produced another hybrid system, a sort of nanorobot that moves thanks to a rat's heart muscle cells. This assemblage, made of a string of silicon in a vaulted arch above which heart fibers are attached, is only as thick as a human hair. Fed by glucose, the robot climbs at a speed of about 40 micrometers per second thanks to the vault's movement of folding and unfolding, induced by the contraction and relaxation of the muscular fibers.

Even if the nanohelicopter has not yet flown, and even if the nanorobot must still be equipped with sensors and a controller to make it actually useful, these research efforts nevertheless open a vast array of applications in the long term, such as pulverizing blockage in an artery or injecting medication into an individual cell. One day it is even planned to replace a deficient gene with a synthetic copy corrected of its imperfections, directly within a cell's nucleus.

13 Hybridizing the Living

Mysterious in the light of day, nature never unveils herself, and there is no lever or machine that can oblige her to make my mind see what she has resolved to hide from it.
—Johann Wolfgang von Goethe (1808)

Although the term *bionics* used as a signifier for a scientific discipline is not yet well known to the general public, it is in vogue as a description of hybrid systems. In two famous TV series of the 1970s and 1980s, the U.S. Secret Service exploited the extraordinary feats that Colonel Steve Austin (with bionic eye, legs, and arm) and Jamie Sommers (with bionic ear, legs, and arm) were capable of performing. Then known under the names the "Six Million Dollar Man" and the "Bionic Woman," such expensive creatures would now be a bargain because the long process of refining a single cyberprosthesis costs almost as much. Ever since a 1960 article in *Astronautics*,[1] people have been using the term *cyborgs*, cybernetic organisms, for organisms equipped with cyberprostheses.

Intelligent Prostheses

Cyberhand
Within a year after the start of its financing, the Cyberhand project had already consumed a million euros. It federated several European teams (coordinated by Paolo Dario of Pisa) with the goal of building an artificial hand capable of reestablishing sensory-motor communication between the hand and the central nervous system. To do so, special electrodes would be implanted to link the spinal nerves of the patient's arm (sensory and motor) to the apparatus. The hand would be controlled by the patient's motor fibers and would send back (via sensory fibers) data that is exteroceptive (such as touch, pressure, and heat) and proprioceptive (such as the

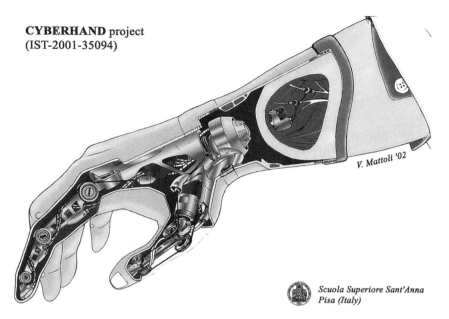

CYBERHAND project
(IST-2001-35094)

V. Mattoli '02

*Scuola Superiore Sant'Anna
Pisa (Italy)*

Figure 13.1
The Cyberhand is capable of sensations. (© Cyberhand project [IST-2001-35094] and
Scuola Superiore Sant'Anna, Pisa.) See color plate 20.

positioning of the fingers and the hand). The apparatus has sixteen degrees
of freedom. It is moved by six micromotors and has a covering with a soft
texture, into which the different sensors are placed with a distribution
similar to that in the human hand—more in the thumb and palm than in
the other fingers. The first refinements were made without the hand's
being connected to a human being, however.

From the motor standpoint, the apparatus has learned to hold fine and
fragile objects such as potato chips, to fill a glass of water, and to grasp a
heavy cylinder. From a sensory standpoint, it is capable of distinguishing
textures and such things as whether an object is too hot or too cold and
whether it is made of metal or of wood. It is these sensations that are the
most lacking in wearers of passive prostheses, and the wearers have trouble
adjusting their movements without this sensory feedback.

The first trials with patients are under way and are proving delicate, for
an electronic circuit must be permanently implanted in their arms. Doing
so will require establishing between the living member and the apparatus
a way of learning the correspondence between two types of instructions,
the bearer's intentions and motor commands. Despite the technical diffi-

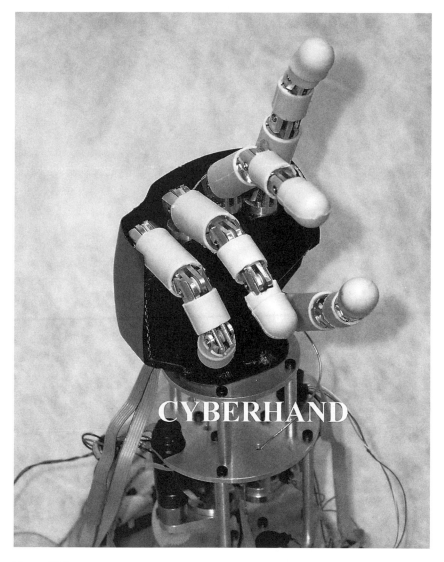

Figure 13.1
(continued)

culties, researchers think they will be able to commercialize Cyberhand by around 2011.

Bionic Leg and Arm

On the same principles, the Canadian company Victhom Bionique Humaine has invented a bionic leg that anticipates the movements desired by the person wearing it. Sequenced movements of horizontal walking combined with those of climbing a staircase are now possible, thanks to a software module of artificial intelligence, whereas a passive prosthesis previously had to be specially calibrated for one or the other of these movements. The current version has autonomous power for six to eight hours of operation by recharging itself thanks to movements of the prosthesis during walking.

The American press has also made much of the case of Claudia Mitchell, a former U.S. marine whose arm was torn off in a motorcycle accident. It took a five-hour operation to implant onto her shoulder the bionic arm perfected by the Rehabilitation Institute of Chicago. This arm, with eight degrees of freedom, responds to impulses given by the patient to her pectoral muscle, to which the surgeons "connected" the nerves of her shoulder. An electronic chip serves to interpret between the contractions of this muscle and the cyberprosthesis. In return, her arm can transmit sensory data when it meets or grips an object. Now Claudia can perform many daily tasks, such as writing, filling a washing machine, cooking, and peeling fruit.

The various wars in the world have unfortunately increased the cases of amputations. This is why DARPA has invested $50 million to improve this type of intelligent prosthesis. According to the head of the program, Geoffrey Ling, it aims for example to reproduce an arm with which "the patient can play the piano." "We're not talking 'Chopsticks,'" Ling adds, "we're talking Brahms!"[2]

Exoskeletons

There are also prostheses in the form of entire suits, American versions of which are oriented toward military and offensive applications, and Japanese versions of which are more designed to help people with disabilities.

The Japanese start-up Cyberdyne—directed by Professor Yushiyuki Sankai of the University of Tsukuba—conceived the Hybrid Assistive Limb, HAL, a suit for aged or handicapped people. The "portable robot" measures 1.60 meters and weighs 23 kilograms and can interpret the muscular elec-

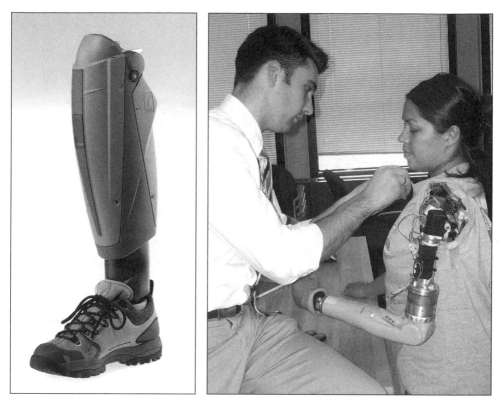

Figure 13.2
Prostheses that adapt to their wearers. Left: Victhom's Artificial A-leg(© Victhom
Bionique Humaine.). Right: Claudia Mitchell's bionic arm. (© Rehabilition Institute
of Chicago.)

trical impulses registered at skin level. It thus can furnish supports for the
legs and arms that augment a person's resistance and stability, facilitating
tasks such as standing up out of a chair, climbing stairs, and carrying up
to 40 kilograms. Thanks to HAL, a Japanese man who had been quadriplegic for twenty years was able to realize his dream of climbing the Breithorn,
a Swiss mountain. Other similar projects are under way, including the
Cyberthosis suit at the EPFL, which programs motor training for disabled
people.

There are also exoskeletons coming from England, Germany, and
France that do not possess electronic chips, but are no less intelligent
applications: they simulate aging! The French company Seniosphère
notably commercializes a British combination called Simulating Age

Figure 13.3
Two exoskeletons, one intelligent and the other . . . clever. Left: HAL, the "intelligent" exoskeleton to improve human strength and stability. (© Yoshiyuki Sankai, CYBERDYNE, Inc., University of Tsukuba. HAL™.) Right: Simulating Age MObility (SAMO) simulates the sensory-motor problems of geriatrics. (© Seniosphère, Paris.)

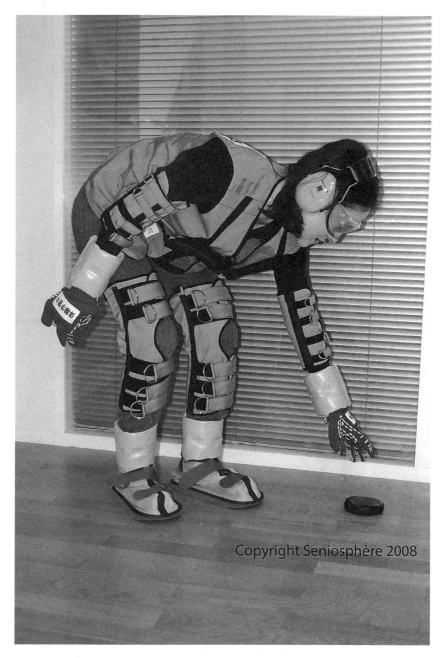

Figure 13.3
(continued)

MObility (SAMO), which allows young people to perceive and understand octogenarians' daily sensations. Its thick tissue equipped with rigid elements weighs down and stiffens its articulations, as is the case with severe arthritis. Glasses and hearing aids also simulate various sensory deficiencies. This combination is used in training programs to sensitize people to geriatric problems; SAMO has become part of the training of home nurses, architects, physical therapists, and supermarket directors. From the start, a company director at Ford ordered it because he wanted his young engineers to understand how to design cars that were comfortable for people older than fifty.

Intelligent Endoprostheses

Sensory Cyberprostheses

When patients suffer from damaged sensory cells or sensory organs but still have intact sensory nerves, then some feeling can be partially reestablished.

After having long implanted encumbering electrodes in the cranium of patients with poor vision, researchers have developed more discrete arrangements. The Argus II system (from the University of Southern California under Mark Humayun) uses a camera mounted on eyeglasses to sent electrical impulses to the optic nerve. This type of implant allows persons suffering from retinal degeneration to apprehend forms, lights, and movements. Perceiving only specks of light at the start of the experiment, volunteers quickly learned to distinguish obstacles in their way and even faces. The bionic eye has already been implanted in twenty or so subjects. A patient who had been suffering from retinal degeneracy for thirty years has already managed, he says, to "sort out white socks, grey socks, and black socks"![3]

Different avenues have been pursued for "seeing" with the ears or the tongue, which requires teaching patients aided by an electronic chip to translate sound signals or weak electrical simulations from the tongue into visual sensations.

Artificial cochleae are more widespread. Contrary to external auditory prostheses that merely amplify sounds, these endoprostheses process sound signals and transmit them directly to the auditory nerve. Therefore, they are reserved for patients who are totally deaf or with severely impaired hearing. A transmitter is worn behind the ear, which sends sound signals to an exterior antenna, which in turn transmits them to the receiver of a subcutaneous antenna. The information is then sent to a set of electrodes connected to the auditory nerve. The problem remaining to be resolved with this type of implant is how to encode and decode satisfactorily the pertinent sounds amid the ambient noise, especially those at frequencies called "conversational"—between 500 and 3,000 hertz—that are characteristic of human speech.

Motor Cyberprostheses

Intelligent electronics is also reaching muscular stimulators. In a classical cardiac stimulator, the intervals between the stimulations sent to the ventricles and those sent to the auricle are fixed, but they ought to adapt to the patient's cardiac rhythm, whether at rest or running, and adapt when a pathology has modified the frequency. The Adapter project, conducted by French labs in Rennes and Strasbourg in partnership with three companies, has a three-year goal of creating an endoprosthesis whose stimulation intervals will vary as a function of the measures continuously taken by new sensors implanted in the subject's heart.

The European program Stand Up and Walk (SUAW)[4] is designed to stimulate the walking movement in patients injured by lesions of the spinal cord: their motor nerves are not usable, but their leg muscles are intact. By electrostimulation, the muscles are directly involved, and an "embedded programming box" links impulses in such a way as to organize various movements. The patient can then push a button to rise and another to walk while balancing himself with crutches or a walker. Researchers are now trying to replace the wires of the implanted electrodes, which have been vectors of infections, with other modes of communication, just as they are planning to miniaturize all their materials.

If the techniques of stimulation do appear beneficial to alleviate handicaps, it must be admitted that some of them do not specifically aim to palliate disabilities, but rather to control animal and human organisms that are in good health! Ethics committees are vigilant in some countries of the world—but only in some.

Radio-Controlled Prostheses

Remotely Guided Animals . . .

Various invertebrates have been transformed into remotely guided missiles. Robo-roach is a bionic cockroach produced by Japanese and Swiss teams. It is equipped with electrodes that send stimulations into the nervous ganglia so that it can move forward or turn to left and right according to a human operator's orders transmitted by a computer.

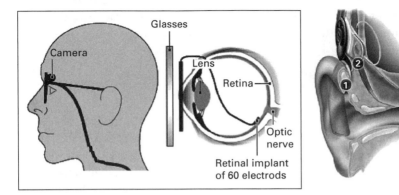

Figure 13.4
Sensory cyberprostheses. Left: The artificial retina Argus II. A camera placed on the lens captures images that are processed and sent to the retinal implant. The electrodes of this implant activate neurons of the optic nerve, which then transmit nervous signals to the visual cortex. Right: The artificial cochlea Digisonic. (1) An external sensor converts sounds into digital signals; (2) a processor sends these signals to an internal implant; (3) the implant converts digital signals into electrical ones, which are sent to the cochlea; (4) electrodes capture these signals and stimulate the auditory nerve like the cells of a natural cochlea, and the patient hears. (© Dunod.)

Researchers at the Robot Engineering Technology Research Centre in Shandong, China, have applied the same approach to a pigeon, with implanted electrodes commanding the pigeon to fly right or left, higher up or lower down.

The arguments justifying such experiments point out the role that these animals equipped with miniature cameras and microphones might play in rescuing the wounded in the event of an accident or catastrophe or in doing civil surveillance. Yet military applications are also envisaged.

. . . and Radio-Controlled Humans!

One might suppose that these devices would be confined to nonhuman organisms, but a Japanese company has already crossed the relevant Rubicon. Nippon Telegraph and Telephone Communication Science labs have perfected a "vestibular galvanic stimulation" device that sends electrical impulses to a human volunteer's inner ear, where the sensory receptors for equilibration are found. A light current at the level of the left or right side makes the subject turn irresistibly in the same direction because a reflex is triggered to recover an equilibrium that he thinks is disturbed.

Figure 13.5
A human controlled by vestibular galvanic stimulation through radio signals! (©
Naohisa Nagaya, University of Electro-Communications, Tokyo.)

Apart from applications that are clearly useful, such as using this device to
remove a pedestrian from the trajectory of a car or to control the equilibrium of old people, other more playful uses are envisaged, such as equip-
ping MP3 players with the ability to trigger dance reflexes, which would
change with each musical tune!

Far from questioning the outcomes of this type of experiment, a
researcher at the Washington University School of Medicine has advocated
the use of electromagnetic fields to act *at a distance* on the vestibular system
and hence on a person not equipped with the previously described device.
He proposed military applications, but also civil ones, such as letting spec-
tators seated in a theater or in a stadium feel sensations analogous to those
that the movements of a dancer or a sports player might engender. To our
knowledge, this project has received no support, either scientific or finan-
cial—as yet.

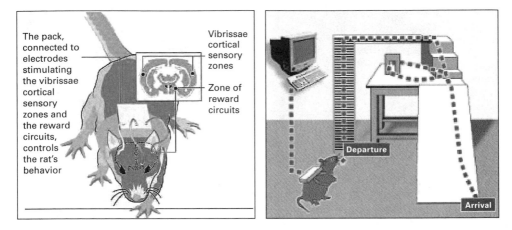

Figure 13.6
A radio-controlled rat learns a predetermined route. Left: A schema of the remote-controlled rat. Right: One of the rat's "forced" routes when it is radio controlled. (© Dunod.)

"Endoprosthetic" Learning?

Researchers under John Chapin at Downstate Medical Center at New York University are also capable of remotely guiding mammals, specifically rats. Unlike the cockroaches or pigeons we mentioned previously that were directly stimulated in their motor zones, these rats undergo a true learning process through reinforcement after stimulation of other zones in their nervous system. They can thereby establish associations between sensory stimuli and specific actions.

To do this, the researchers place some fine electrodes in cortical sensory zones corresponding to the rat's left and right vibrissae, and other electrodes in the telencephalon circuits of "reward," which when stimulated procure an agreeable sensation for the animal. The rat has to learn to turn to the left (or right) when the left (or right) vibrissae are stimulated or to go straight ahead when they both are stimulated at once. The rat gets the reward stimulation at each correct behavior.

The animal quickly learns to associate stimulations of the vibrissae with the appropriate locomotion. Equipped with a harness containing a microstimulator controlled at a distance by a computer, the rat can then be teleguided to follow trajectories that it would otherwise not take naturally because they lead, for example, to climbing high stairs or traversing strongly illuminated areas.

The rat is then trained in the same way to make a ten-second pause when it detects the odor of an explosive. The researchers foresee using these rats, just like other radio-controlled animals, for rescuing people or detecting mines; they say that despite the cost of equipment, this method is less expensive than using a robot specially trained for this kind of mission.

Stimulating and Recording Endoprostheses

The preceding researchers do not have access to the naturally produced signals that are processed by the rat's olfactory nervous system. Other American laboratories have carried out experiments in which the subjects are controlled at the same time as their sensations are recorded. Thus, DARPA has financed research at Boston University to implant recording electrodes in the sensory circuits of small sharks a meter long—the circuits that allow them to detect the electrical fields and the smell of the prey they are pursuing. A microprocessor translates these data into information useful to a human, such as the proximity of fish, mines, or submarines.

Other electrodes, not recording but stimulating, might be implanted in the sharks to give them a false sensation of the presence of prey and thus serve to pilot them toward predetermined targets. DARPA quickly classified this research program as a defense secret, for the silent and energy-autonomous sharks might be much more effective spies than robots. It is now being pursued at the Naval Undersea Warfare Center at Newport, Rhode Island, along with more fundamental research aiming to understand the sensory processing that these animals perform in nature, especially the way they use the earth's magnetic field to orient themselves. This research might also serve to detect banks of fish or changes in ocean temperature or even to chase away large sharks from the coasts of Florida.

Double Agents among the Cockroaches

Endoprostheses would therefore be able to manipulate an animal's behavior. One very original experiment shows that the equivalent exists not only at the level of an organism, but also at that of a whole group. Indeed, it demonstrates that one can manipulate the collective behavior of a colony of cockroaches by "implanting" into its midst a few robotic agents. This is the goal of the European Leurre ("lure") project coordinated by Jean-Louis Deneubourg, which is studying the possibility of constituting mixed groups in which animals and robots interact.

Figure 13.7
The Leurre project. One of the Insbots that have trained cockroaches to prefer a particular shady zone. (© Projet Leurre, Université Libre de Bruxelles et École polytechnique fédérale de Lausanne.) See color plate 18.

Miniature robots called "InsBots" (Insectlike robots) do not resemble their future companions, however. These cubes with wheels possess infrared sensors and a light sensor to detect obstacles and surrounding insects, as well as to identify the dark zones in which cockroaches have the reflex to hide. They can detect insects' olfactory signals and emit the pheromone that indicates group membership and serves for internal communication.

The researchers first determined the principal characteristics of cockroach behavior, such as the random aspects of their movements, the frequency of their stops, the probability of their remaining alongside their fellows, and so on. The InsBots were then programmed to exhibit similar behavior so as to move in a nonaggressive way amid the group and to react in a manner suitable to the animals' chemical signals.

In this way, the researchers were able to provoke in a noncoercive way behaviors that were not observed in the absence of these robots, such as modifying the cockroaches' collective choice of shelter. In effect, when an isolated cockroach has the choice between several shadow zones that are more or less comfortable, it chooses one of the zones at random. Then due to chemical communication among these insects, all the cockroaches move little by little toward the most "welcoming" zone. The InsBots have managed to manipulate this communication in order to entrain the insects toward a particular shadow zone. Researchers have also observed that the group of cockroaches has the tendency to keep the same shelter even when the InsBots are no longer there.

Apart from the importance of this research for understanding the self-organization of insect societies, various applications are foreseen. In particular, campaigns of eradication might rely on the robots' capacity to lead cockroaches as a group toward a site where they might be eliminated, like the Pied Piper did with rats or Virgil did with Neapolitan flies.

Endoprostheses for Identification

Let us go back to more classic examples that further illustrate the vast field of endoprosthetic applications.

Like a car marked with an antitheft engraving, the human body might soon bear a lifelong identification mark. The American-Canadian firm Verichips sells implantable electronic chips the size of a grain of rice that can contain data on an individual identified by a sixteen-figure number. The chips are inserted under the skin in a painless way and present no risk of infection. The U.S. Food and Drug Administration has permitted their use in reading only, but other countries—notably in South America—allow both reading and writing chips. Most of these chips are to hold medical files, accessible even if the bearer is unconscious. Others are provided with systems of radio frequencies, like GPS systems that detect where a child or an aged person can be found, or else systems of biometric detection, sort implanted badges that must be presented to enter into sensitive sites that are not authorized for everybody. It is the latter use that the attorney general of Mexico and his team have taken up to control access to secret documents in the war against the drug cartels.

With regard to microchips, we cannot ignore the case of Kevin Warwick, professor of cybernetics at Reading University in England, who had the odd idea of implanting in his forearm a glass capsule that emitted signals and served as a remote to open a door, put on the lights, or move a robotic hand at a distance. He even claims to have succeeded in the first experiment of "electronic communication" with another human being—his wife—herself implanted with a similar device, but it seems that the results were much more modest than predicted!

Other studies have a long-term but more serious (and more useful) objective, such as using this type of endoprosthesis to deliver medication on demand and at a distance to patients, which would revolutionize the therapy for diabetics, for example.

"Mind over Matter"

Some of the preceding research creates an uneasy feeling, however, because the altruistic applications envisaged for remotely controlled subjects are easily convertible into less well-intentioned applications. Moreover, persons carrying the identifying chips are not safe from the possibility of misappropriation or falsification of the registered data.

The studies that follow are of a quite different order because in them the control is reversed: it is not electronics that master subjects, but nervous impulses that govern the machine.

The principle is to record in real time, with multielectrodes called "neuroprostheses," the electrical activity of vast populations of sensory and motor neurons and to translate the corresponding data into motor orders capable of activating a machine—a computer or a robot. Several recording techniques are used to realize this feat. Invasive or semi-invasive techniques record nervous electrical signals with the help of neuroprostheses implanted under the cranium or directly in certain zones of the cortex. Noninvasive techniques record an electroencephalogram (EEG), as during a medical procedure, by the intermediary of a bonnet covered with many electrodes.

An enormous budget has been given to these experiments, which are contributing to the refinement of what is called *brain-computer interfaces* (BCIs) or *brain-machine interfaces* (BMIs) and which have the vocation of improving the quality of life for people who are severely handicapped at the motor level.

Invasive Neuroprostheses

The experiments with invasive neuroprostheses are not recent; Eberhard Fetz at Washington University began them in the 1970s and William Grey Walter, the father of the cybernetic turtles and the pioneer of the EEG, preceded them with a little known experiment in 1964. In this experiment, called the "precognitive carousel," he used a few subjects who were already equipped with recording electrodes in their motor cortex for health reasons and gave them a remote control supposedly for a carousel slide projector. They were not told that this remote control was not connected to anything and that only their amplified EEG signals were being directly sent to the slide projector! The subjects very quickly became very uneasy, for the slides began showing *before* they had the time to press the button on the remote control, at the very moment when they only *thought* of doing it!

It took twenty-odd more years for the corresponding techniques to prove really operational. In effect, the neuroprostheses had to have been accepted in the nervous tissue for a long period so that the many electrodes could record simultaneously the large population of neurons (more than one thousand units) in various cortical zones. The surgical risks of such operations were far from minor and had to be carefully evaluated. These devices had to send signals that were sufficiently precise to be able to be correctly processed. Moreover, the sophisticated algorithms had to be refined in order to execute the processing and to categorize nervous activities in a reliable way so that they would correspond to the movements the subject wanted to perform—and, better still, would anticipate them.

Several labs are working on these technical and methodological issues. The projects Direct Brain Interface, conducted by American and Austrian teams, and BrainGate at the Cyberkinetics Neurotechnology Company, have the goal of controlling through thought certain mobile apparatuses and devices for vocal and written communication. The subjects recruited suffer from either severe lesions in the spinal cord or muscular dystrophy or Locked-in Syndrome (LIS), in which the patient is imprisoned in his body, with very few signs of motor communication. Although this research has undeniably improved these patients' lives, it can as yet order only very simple motor actions, such as triggering a switch, turning up the television volume, moving a robotic hand.

The most advanced work up to now is that of Miguel Nicolelis, professor of neurobiology at the Center for Neuro-engineering at Duke University and in Switzerland, who has experimented with invasive neuroprostheses on rats and then on monkeys. In his laboratory, Rhesus monkeys have learned to control the movements of a roboticized arm so that the arm moves toward a precise point; the gripper situated at its end is somewhat prehensile. The electrical signals of hundreds of neurons gathered by multielectrodes implanted in the monkeys' various cortical zones (parietal and frontal, corresponding to sensory and motor) are used.

In a learning phase, a monkey acts on a joystick to direct a screen cursor toward luminous circles, whose position and size inform it about the distance of the gripper to the target as well as about the force of prehension that the gripper is exerting. When the monkey obtains the desired composition of movements, it is rewarded by the delivery of a fruit juice. At the same time, a computer analyzes the signals gathered by the electrodes, categorizes them, and learns which electrical patterns are associated with which movements of the cursor and the robotic arm.

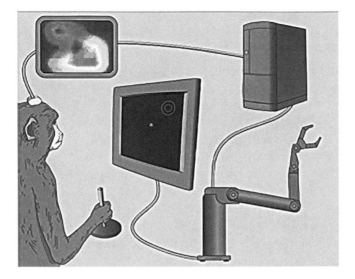

Figure 13.8
One of the brain-computer interface experiments with invasive neuroprostheses in
a monkey. The monkey's brain activity, corresponding to the movements of its arm
when it uses the joystick to move a cursor toward a target, is recorded on the computer. This activity is translated into electrical impulses that move a robot's arm in
the same way as the monkey moved its arm. After training, the monkey needs only
to "think" of moving its arm for the cursor to move toward the target—and hence
for the robot's arm to execute the movements that the monkey has not actually
performed itself. (© Dunod).

After this learning phase, the joystick is removed. It has been noticed
that some monkeys continue to move their arm as before, but in the
void, for they "realize" that they can still direct the cursor and the robotic
arm. It is obviously the computer, continuing to analyze their neural
activities, that provokes the corresponding movement when it recognizes
a learned electrical pattern. After several trials, it even occurs that some
monkeys no longer make any movement at all, but manage to guide the
robotic device only by "force of thought" in order to obtain their reward.
Cerebral imagery analysis of the motor cortical zones corresponding to
their own body representation shows the remarkable fact that these
monkeys have integrated the artificial arm as a natural extension of their
own organism.

Analogous research has recently allowed a human to control the walking
of the android robot ASIMO by the same procedure.

Noninvasive Neuroprostheses

According to specialists, the future of invasive neuroprostheses is uncertain, if only because of the ethical problems that they raise. Some have turned to other techniques, such as those using noninvasive neuroprostheses, specifically bonnets with multiple electrodes that record subjects' EEGs. Such a procedure is interesting: nothing prevents its being used on a consenting subject, but it does pose the problem of more difficult processing because the signals that can be exploited are considerably weaker. For the moment, it seems that these noninvasive neuroprostheses enable very simple motor commands, but not any control of the complex movements of machines that have several degrees of freedom.

Among the places where this research is being conducted are the Brain-Lab at the University of Georgia and the Fraunhofer Institute in Berlin. In the former, the researchers are perfecting arrangements to move an electric wheelchair or to enable mental surfing of the Internet (as of today, in a very basic way). At the latter lab, researchers are devising a "mental typewriter," Hex-o-Spell, with which a subject wearing a bonnet of 128 electrodes chooses letter by letter on a computer screen the word he wants to spell. This choice is made by stages: the subject points first to a particular group of letters, then to the chosen letter in this group. The brain wave on which these choices rely is the "P300"—a positive cortical activation that is triggered about 300 milliseconds after an expected stimulus is detected. Thanks to this arrangement, the subject can currently write almost ten words a minute.

Other labs participate in a European project called Adaptive Brain Interface (ABI) at the Dalle Molle Institute for Perceptual Artificial Intelligence in Switzerland. This work is specifically focused on the search for algorithms able to process multimodal neuronal data. It already enables a human subject to direct a small wheeled robot at a distance, mentally ordering it to perform simple actions such as turning to the left or right or moving straight ahead.

A professor at Tubingen University, Niels Birhaumer, has obtained perhaps the most original results. With the money he received for having won the Leibniz Prize for previous work, he financed his research on a system named the Thought Translation Device, thanks to which a patient with LIS was able after a month's training to compose messages on a virtual keyboard. With this procedure, the patient sent a word of thanks to the professor through the intermediary of the journal *Nature*, and he even explained his method of mental translation in an interview in the *New Scientist*!

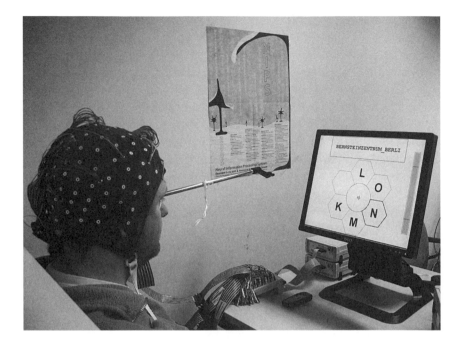

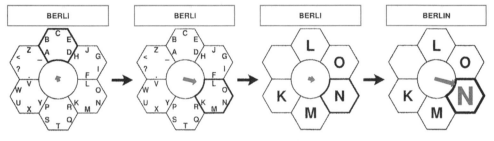

Figure 13.9
The "mental writing machine." Top: A noninvasive neuroprostheses: the arrangement of the mental writing machine Hex-o-Spell. Bottom: The different stages of choosing a letter of the alphabet by thought using this arrangement. The subject chooses mentally more and more detailed zones in which the letter he is thinking of is found. (© Klaus Müller, Technical University of Berlin.)

Figure 13.10
An experiment of locomotion "controlled mentally" in a virtual reality environment at the Graz University of Technology. (© Robert Leeb, Graz University of Technology.) See color plate 21.

Finally, we mention a European project coordinated by the Austrian laboratory of the Technology University in Graz in which researchers use virtual reality to analyze more precisely the interactions between the subjects and an environment in three dimensions that the researchers can perfectly control. In one of the experiments, a subject was trained to advance or stop in a virtual street lined with shops by imagining that he was performing an agreed movement, such as moving a foot or a hand. He could even engage in conversation with avatars (characters representing the players in a virtual game) met along the way, provided that he managed to stop at a suitable distance for communication.

The progress of these various procedures of noninvasive neuroprostheses is promising, yet it cannot obscure all the residual difficulties. The corresponding learning periods can be long and arduous, all the subjects do not necessarily succeed at them, and the controls are still too simple for massive applications to be hoped for in the short term.

It is no doubt for these reasons that "biosignals" other than neuronal activations—for example, the signals registered in the muscles, on the skin, or in the flow of blood in the brain—are also the subjects of much research. At NASA's Ames Research Center in California, researchers are working on subvocal movements. With three recording electrodes placed in the throat, the subject can pronounce audible words without speaking aloud. A program translates into sound or into text the muscular microcontractions coming from the larynx and the muscles of the tongue when the subject reads each word to himself. The system is not yet perfect, for it can compose only twenty-odd words and numbers from zero to nine.

Its inventors envisage an entirely portable apparatus, which would suit persons working in a noisy environment or people wanting to have a private conversation over the cell phone—also a dream for their neighbors in public places!

Conclusion

It is sad to think that nature is speaking and the human race is not listening.
—Victor Hugo (1870)

These many excursions into the achievements associated with innovations inspired by biology demonstrate how this field of research is lively all over the world. But this should not obscure either the major and still intractable limitations of the work described in each of the three parts of this book or the ethical questions that the work poses.

Biology and Technology

The technological inventions inspired by nature are enjoying an astonishing takeoff, due notably to advances in nanotechnology. However, it is true that the efficacy of most natural structures, processes, and materials rests above all on their organization at the molecular level. Even if humankind can manage to imitate them at this level (still to a very limited extent), it would not explain how the structure and composition of many other levels in the hierarchy are organized: organelles, cells, tissues, organs, and so on—which are the key to the remarkable functions of nature's products. Some researchers, including Julian Vincent of Bath's Centre for Biomimetics and Natural Technologies, think that although we probably have the tools to imitate some biological inventions as well as is possible, we still have a hard time approaching their perfection, for we devise them with fundamentally different strategies.

For example, whereas we constantly seek to invent new basic materials, including about 350 types of synthetic polymers, nature uses only two materials, proteins and polysaccharides. Whereas we seek to fabricate smooth surfaces and perfectly symmetrical structures, nature favors rough and hole-filled surfaces of only approximate symmetry. Whereas we generally look for an optimal solution to a single problem, nature prefers solutions that are just good enough to resolve several problems at the same time.

Nature possesses many other qualities from which we ought to take inspiration, like those cited by Janine M. Benyus in this "hymn" taken from her best-selling book *Biomimicry: Innovations Inspired by Nature*:[1]

Nature runs on sunlight
Nature uses only the energy it needs
Nature fits form to function
Nature recycles everything
Nature rewards cooperation
Nature banks on diversity
Nature demands local expertise
Nature curbs excesses from within
Nature taps the power of limits

It is the first two qualities that most distinguish the productions of engineers from those of nature. Whereas we draw abundantly on fossil energy resources, life maintains itself effectively thanks to solar energy. Whereas we still worry too little about our energy expenditure—because it is only by increasing it that we resolve most of our technical problems— living organisms have evolved by economizing energy at all levels, structural and functional.

Attentive study of the way nature achieves this economy while it resolves problems much more complex than our own is necessarily profitable in these times when our species has finally realized that sustainable development is a priority.

Biology and Robotics

Bioinspired animats and robots are still far from attaining their living models' decision-making autonomy.[2] Nevertheless, they are enjoying growing success in various applications that involve environments or tasks that humankind cannot characterize very well. They can act to explore distant planets, to intervene in inhospitable sites to rescue or to repair, or even to lend automation to home help. Nor should we neglect to mention the potential for playful applications: there are many interactive robots that play with children (and adults), and we are starting to create entertainment in which robots and live comedians trade jokes.

Other applications might have more debatable consequences. This is why several countries have drawn up regulations guaranteeing the ethics of research on robots. After the first International Symposium of Robo-ethics (organized in 2004 in Italy), the European Robots Research Network (EURON) in 2005 established a discussion group composed of researchers entrusted with establishing guidelines for legislators.[3] Participants referred to several older and more recent texts, such as the Universal Declaration of Human Rights (United Nations 1948), the Manual for Scientists and Researchers (Organization for Economic Cooperation and Development, 1994), the Declaration on the Sciences and the Use of Scientific Knowledge (UNESCO, 1999), and the Charter of Fundamental Rights of the European Union (European Parliament, 2000). Judging it premature to rule on the eventual emergence of human specificities among machines (such as con-science, free will, or a sense of dignity), they preferred quite reasonably to draw up a list of the problems and benefits associated with existing robots and then to put forth recommendations targeting various categories of creatures such as humanoids, industrial robots, domestic robots, explora-tion robots, assistance robots, military robots, virtual robots, and educa-tional and entertainment robots. A general point on which they particularly insisted was the need to promote popular discussion to increase public attention to the problems of robo-ethics.

In 2004, Japan adopted the Guide for the Security of Performance of the Next Generation of Robots. Similarly, South Korea developed its own charter in 2007, for this country is pursuing an ambitious program: to introduce into each home between now and 2020 a robotic companion that will know how to entertain, teach, protect, and help in domestic tasks. To prevent harmful uses, its minister of commerce, industry, and energy has assembled a committee made up of robotics experts and scientific and literary futurologists to decide on the roles and functions of robots "expected to develop strong intelligence in the near future."[4] Except for a modern adaptation of Isaac Asimov's famous three rules of robotics[5] and the requirement that each machine's decisions and actions should be trace-able, the sages wanted to include in this charter a few articles to prevent various misuses. It appeared that some cases had already arisen, such as the requested authorization for mixed marriages between robots and humans!

Do all robots have the same chance of being asked to marry? Or (a more serious question) do humans feel the same emotions when they interact with any humanoid? The answer is not simple. In an article published in 1970, the Japanese roboticist Masahiro Mori described what he called "the uncanny valley" ("Bukimi no tani"), a Freudian concept suggesting that because a robot is both familiar and strange to us, it provokes an intense sensation of malaise. Here the term refers to a theoretical curve that exhibits a dip that represents a plunge in our sympathy toward a robot when it has an "almost human" appearance. On the contrary, when it is clearly distinct from us (like the robots Aibo and Kismet) or when it is clearly close to us (like Repliee Q2), our feelings are much more positive.

This curious theory has found an explanation from some biologists, who put forward the hypothesis that our brain has evolved so as to detect in a human appearance the elements that indicate a person "in good health." An almost human robot would therefore send signs of abnormality that would displease observers, such as the case of an android that did not have eyes or express any emotion. On the other hand, other researchers criticize this theory (and this explanation) for being pseudoscientific, arguing that it has never been studied in a controlled way and that a whole gamut of emotions has apparently been detected in all sorts of machines, whether of humanoid appearance or not! Nevertheless, those who devise the nonplaying characters in video games or machines that interact with people do appear to give some credence to this theory.

Other scientists stress that the eminently cultural aspect of our relations with robots refers back to our concept of humanity. For Westerners, a new robot that increasingly resembles a human will dangerously approach the frontier that it is forbidden to cross, which still ensures human superiority over machines. For Asians, such a robot is designed

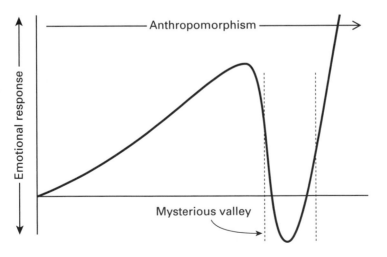

Figure 14.1
The curve of the "uncanny valley." (© Dunod.)

to occupy quite simply its place *as* a robot and to perform its functions as a *robot* as well as possible (as well as or better than a human). These opposite philosophies are fed on both side of the Bosphorus by artistic and literary products that merely reinforce them—the frightening Golem and the creatures Frankenstein and Metropolis, on the one hand, versus the sympathetic Tetsuwan Atomu (Astro Boy), Mazinger Z, and Tamagochi, on the other.

This is why Asian robots are usually designed to become true companions for elderly or disabled persons, whereas in the West they are confined to executing useful tasks without making any particular bonds with us. In short, the future Western domestic robots will always use the service stairs.

Another reason for this difference is the little good press that the West gives to robots, which means that animats (those animal robots whose autonomy we are trying to increase) are still generating fear. Would they really become dangerous if they one day took control of their own destiny? Probably yes, if they are conceived with a view to hurting, and probably no, if their role is to back up humans in difficult or dangerous tasks. This ambivalent response is in fact valid for any applied scientific research: some radioactive substances are sources of benefits in nuclear medicine, but they can also cause millions of deaths; bacterial cultures in a laboratory are used for gene therapy, but they can also be spread across a country for massive extermination of its inhabitants. If robotics inspired by biology does not carry any less risk (but also no more) than does applied research in other scientific fields, we don't want to indulge the few pseudoscientists who predict in the medium term—for sensational publicity—the destruction of the human species by artificial creatures.

Organic and Inorganic

Research on living-artificial hybrids is in full expansion but has not given rise to any new charter, for it depends on the same laws as those governing experimentation on the living. Voices call for strengthening the rules, in particular to guarantee the longevity and efficacy of hybridization, whether this means increasing the capacities of a robot or of a living organism, human or not. A recent French book, Joël de Rosnay's *2020: Scenarios of the Future*,[6] points to an ethical gap between "repaired" or "transformed" humans, on the one hand, and "augmented" humans on the other, with the latter considered as testifying to a new will to go beyond human norms by electronic and technical means. In time, this would risk engendering two classes, superhumans and subhumans, divided by the money they can invest in futuristic equipment.

This dilemma has already been confronted in the case of Oscar Pistorius, a South African athlete born without tibia bones. His legs were amputated when he was eleven months old, so he has two passive J-shaped blade-footed prostheses made of carbon fiber, from which he was given the nickname "Bladerunner." After having done spectacularly well in races for the disabled, he managed to finish second in the regular South African championships in the 400 meters and so applied for the right to compete in the Olympic Games in Beijing in 2008. At first, the powerful International Association of Athletics Federations refused to accept him, relying on expert medical opinion that he would have an advantage due to his prostheses and invoking Article 144-2 of the rules that stipulate that no athlete has the right to use "any technical device, including springs, gears or any other element that confers an advantage in relation to those athletes who do not use them." Later on, the federation's decision was reversed by a Court of Arbitration for Sport ruling that the federation had not provided sufficient evidence to prove that Pistorius's prostheses gave him an advantage over able-bodied athletes. Unfortunately, having been thus authorized to compete, Pistorius didn't manage to meet the qualifying times, and the Olympics took place without him. He nevertheless participated in the Paralympics and won three gold medals in the 100-meter, 200-meter, and 400-meter sprints. Clearly, this unprecedented case does raise the formidable expectations that hybrid techniques hold for the disabled.

Progress might be just as significant in the realm of electronic nanochips, which may compensate for sensory or motor deficits, monitor the state of our internal organs, or dispense medication advisedly. Most of the chips are currently between 10 and 100 millionth of a meter (micrometer) in size—in comparison, the average diameter of our red blood cells is 7 micrometers—and other nanochips of a billionth of a meter (nanometer) have just been realized. Nevertheless, the most spectacular revolution lies in the choice of the materials of which they are made—silicon being replaced by substances better tolerated by the organism—and in their mode of construction, the self-assembly of living molecules.

We may suppose that the work of ethics committees will bear on the boundary between an organic person and a mechatronic one. Should this boundary be defined quantitatively—that is, by the ratio of flesh to prostheses, of carbon to silicon? Or qualitatively, by asking when the human being's integrity and personality are threatened by "one too many neuroprostheses"?

Prospects

Rodney Brooks, one of the pioneers of bioinspired robotics and already mentioned several times in this book, recently proclaimed four fundamental objectives to attain in the future to increase robots' autonomy: first, the ability to recognize objects that is equivalent to a two-year-old child's, then the language comprehension of a four-year-old child, then the manual dexterity of a six-year-old, and finally the social comprehension of an eight-year-old child. To someone who asked him how long it would take to attain these objectives, he replied: "Ah. You must be a reporter. I'll never answer that because, you know, in 1966 they thought it was going to be three months for object recognition."[7] Predictions of the same kind were made by the founders of artificial intelligence, assuring us in 1956 that the reproduction of human intelligence was a matter of only a few years away.

Brook's witticism really applies to any bioinspired research. What appears simple, obvious, and clever to us in nature's production in fact has been patiently elaborated for a little more than three billion years. With our ingenious brain, is *Homo sapiens sapiens*, which as such is only a hundred thousand years old and uses techniques that have progressed only over a few decades, sufficiently armed intellectually and practically to attain the degree of perfection observed in the natural world? Will our representation of the world, with which we reason and which is closely linked to our particular sensory-motor equipment, enable us to tackle these problems? And above all, will the goals fixed in a given temporal context still be valid a few years hence?

Without taking these reservations into account, some researchers in the futurology department of British Telecom have projected themselves into the near future—that is to say, fifty years from now. Here are some of the advances they promise us during the coming decades:

• Between 2006 and 2010, toys will express emotions and will recognize emotions in their owners;
• Between 2008 and 2012, sensitive tissues will measure our cardiac frequency and our temperature and estimate our mood, and automatically adjust the lighting and temperature of our home;
• Between 2011 and 2015, an automatically piloted car will use satellite navigation and will have sensors able to stop in case of imminent accident;

• Between 2013 and 2017, robots will replace seeing-eye dogs;
• Between 2016 and 2020, rights concerning the protection of hybridized robots with natural organs will be instituted; the use of devices for controlling emotions—by suppressing anger and stimulating positive sentiments—will contribute to making criminal activities disappear;
• Between 2031 and 2035, computers will become more intelligent than humans (we note that the researchers do not specify *which* intelligence is at issue!);
• Beyond 2051, our thoughts, feelings, and memory will be transferable to a computer, thereby ensuring a kind of digital immortality and the possibility of downloading this information into another human being.

To this list can be added the following prediction made by Daniel Pauly, director of the Fisheries Center at the University of British Columbia, one of the "fifty most brilliant scientists in the world," who was asked to participate in the same exercise in futurology in the fiftieth anniversary issue of *New Scientist*: "A device that could detect, amplify and transmit to us the emotions and fleeting, inarticulate 'thoughts' of animals in such a form as to evoke analogous emotions and thoughts in human brains. This would first work with primates, then mammals in general, then the other vertebrates including fish. This would cause, obviously, a global revulsion at eating flesh of all kinds, and we would all become vegetarians"[8]

This presupposes, of course, that a device claiming to translate the emotions of plants never comes around!

Epilogue

"In order to question there must be two of you—the one who questions, and the one who is questioned. Intimately bound up with nature, the animal cannot question it. . . . The animal is *one* with nature, while man and nature make *two*. . . . De-natured animals, that's what we are," wrote the French novelist and naturalist Vercors more than a half-century ago.[1]

Thus, questioning nature is the prerogative of our species. We hope that this book encourages humans not to pass up this opportunity.

Notes

Introduction

1. The coinage of the word *bionics* is attributed to Major Jack E. Steele of the medical aerospace division of the U.S. Air Force, which organized this conference.

Chapter 1

1. Olivier Razac, *Barbed Wire: A Political History* (New York: New Press, 2002).

2. In the "Cambrian explosion" of around 550 million years ago.

Chapter 2

1. The more this coefficient approaches 1.0, the more a car resists advancing in air: for example, a 2 CV (an old Citroën) has a coefficient of 0.52, and a Mercedes-Benz S320 CDI has a coefficient of 0.26.

Chapter 3

1. A collaboration has also begun with Constantino Creton, a researcher at the École supérieure de physique et de chimie industrielles in Paris.

2. Researchers at the University of Illinois in Evanston recently produced "geckel" (from the terms *gecko* and *mussel*), used to cover repositionable adhesive bands under water, which adds to the nanostructures of the gecko method an adhesive that resists water. We discuss this adhesive in the next chapter: a glue made by the blue mussel.

3. Serge Berthier, *Iridescences: The Physical Color of Insects* (Paris: Springer, 2006).

4. From the Greek *arguros*, "silver," and *nein*, "weave."

5. Jean-Henri Fabre, *Souvenirs entomologiques: Etudes sur l'instinct et les mœurs des insectes,* 9th series (Paris: Delagrave, 1905), 167.

Chapter 4

1. Saint Hilaire describes the product in a dissertation titled "Spider Containing the Virtue and Properties of This Insect, with the Quality and Use of the Silk That It Produces and the Drops Taken from Them for the Healing of Apoplexy, Lethargy, and All Soporific Maladies."

2. Otherwise, we would be able to bend our backs 720,000 times per hour without getting tired!

Chapter 5

1. "The invention that tradition attributes to the Pythagorian Archytas should not surprise us, even if might appear frivolous. Most of the best-known Greek authors and the philosopher Favorinus, great lover of antiquities, report very officially that a wooden dove was constructed by Archytas according to certain calculations and mechanical principles, and that it flew. It was probably by a system of counterweights that it was held in the air, and by the air pressure contained inside that it moved forward. Permit me to cite Favorinus himself on one scarcely believable fact: 'Archytas of Tarentus, both philosopher and mechanic, fabricated a dove of wood that flew, but once it landed, it could not take off again." Aulu-Gelle, *Petites histoires des grands hommes: Les Nuits Attiques*, trans. Fabrice Emong (Paris: Arléa, 2006), book X, section XII, 8.

2. The main title of this essay is "Le mécanisme du flûteur automate" (Paris: J. Guérin, 1738).

3. Alfred Chapuis and Edouard Gélis, *Le monde des automates* (Geneva: Editions Slatkine, 1984; 1st ed., 1928), 151; see also Alfred Chapuis, *Automata: A Historical and Technological Study*, trans. Alec Reid (New York: Central Books, 1958).

4. In Armand Mattelart, *The Information Society: An Introduction* (London: Sage Publication, 2003), 33.

5. They can still be seen in action at the Museum of Art and History in Neuchatel, Switzerland.

6. In Alfred Chapuis and Edouard Gélis, *Le monde des automates* (Geneva: Editions Slatkine, 1984), 276.

7. Clark Hull, *The Principles of Behavior* (New York: Appleton-Century, 1943), 29–30.

8. Ibid., 27.

9. Biologists name the sensory receptors that deal with external and internal stimuli, respectively, "exteroceptive" (dealing with information on the external environ-

ment) and "proprioceptive" and "interoceptive" (dealing with bodily movements and the state of the viscera).

10. In Roberto Cordeschi, *The Discovery of the Artificial: Behavior, Mind, and Machines Before and Beyond Cybernetics* (Dordrecht: Kluwer Academic, 2002), xv.

11. The traces of protorobots were fortunately conserved thanks to Robert Cordeschi, *The Discovery of the Artificial: Behavior, Mind, and Machines before and beyond Cybernetics* (Boston: Kluwer, 2002).

12. Norbert Wiener, *Cybernetics, or Control and Communication in the Animal and the Machine* (Cambridge, MA: MIT Press, 1948).

13. Albert Ducrocq, *L'ère des robots* (Paris: Julliard, 1953).

14. In Albert Ducrocq, *A la recherche d'une vie sur Mars* (Paris: Flammarion, 1976), 244.

15. In John McCarthy, Marvin Minsky, Nathaniel Rochester, and Claude Shannon, *A Proposal for the Dartmouth Summer Research Project on Artificial Intelligence* (1955), 1 (typescript housed in the archives at Dartmouth College and Stanford University).

16. Allen Newell and Herbert Simon, "GPS: A Program That Simulates Human Thought." In H. Billings (ed.), *Lernende automaten* (Munchen: R. Oldenbourg, 1961), 199–124.

17. The problem of the Hanoi towers is a thinking game imagined by the French mathematician Edouard Lucas. It consists of moving disks of different diameters from a tower of "departure" to a tower of "arrival" by passing through an "intermediate" tower in a minimum of moves, while respecting the following rules: (1) one cannot move more than one disk at a time, (2) one can place a disk only on top of a larger one or on an empty place—it being presumed that this latter rule is also respected in the opening configuration.

18. Robert Lindsay, Bruce Buchanan, Edward Feigenbaum, and Joshua Lederberg, *Applications of Artificial Intelligence for Organic Chemistry: The DENDRAL Project* (New York: McGraw Hill, 1980); Bruce Buchanan and Edward Shortliffe, eds., *Rule-Based Expert Systems: The MYCIN Experiments of the Stanford Heuristic Programming Project* (Reading, MA: Addison-Wesley, 1984).

19. The 1972 edition was followed by Herbert Dreyfus, *What Computers Still Can't Do* (Cambridge, MA: MIT Press, 1992).

20. Alan Turing had in 1950 proposed a test permitting the evaluation of a computer's "intelligence" by comparing blindly the responses of a computer program and those of a human brain given by the intermediary of a screen. If an observer could not manage to divine which was the natural interlocutor and which was the artificial one, the computer was considered "intelligent." This test is still used today,

and a reward, the "Loebner Prize," is given each year to the computer program that best manages to fool an observer in a given lapse of time. In fact, no program has yet managed to fool a human in the long term.

21. These researchers' first international conference, "Simulation of Adaptive Behavior (SAB): From Animals to Animats," was organized in Paris in 1990. Thereafter, it has been held every two years, alternating between Europe and other countries of the world.

22. Michael Gross, *Life on the Edge: Amazing Creatures Surviving in Extreme Environments* (New York: Plenum Trade, 1998).

Chapter 7

1. In other terms, a fly would perceive a succession of images as discontinuous if they went by at less than three hundred per second. In humans, this number is between twenty-five and thirty images per second.

2. Many insects and a few birds perceive polarized light. The Vikings used cordierite for this purpose, a stone that allowed them to reckon the position of the sun by observing the stone's changes in color. They turned the stone until it became lighter to obtain the direction of the sun in cloudy weather. Thanks to it, they were able to navigate to the New World and sail back long before Christopher Columbus did.

3. A dozen European teams were involved in the Integrating Cognition, Emotion, and Autonomy (ICEA) project. Psikharpax is the name of the king of the rats in a parody of the *Iliad* falsely attributed to Homer: the *Batrachomyomachy*.

Chapter 9

1. "The single most important unreported story of the decade," according to V. S. Ramachandran, director of the Center for Brain and Cognition (University of California), who compares it to the discovery of DNA. It opens up prospects for studies of language, cultural transmission, and social interactions.

2. Located in 1971, largely in the hippocampus.

3. Discovered in 1984, in the postsubiculum, the retrospenial cortex, and other structures, especially thalamic.

4. Discovered in 2005 in the entorhinal median cortex.

Chapter 10

1. Antonio Damasio, *Descartes' Error: Emotion, Reason, and the Human Brain* (New York: Avon Book, 1994).

2. Excerpt of the title of John Holland's article in *Scientific American* (July 1992), 66: "Genetic Algorithms." Computer program that "evolve" in ways that resemble natural selection can solve complex problems even their creators do not fully understand.

3. The name is taken from the pilot of an aircraft called the *Albatross* in the Jules Verne novel *Robur the Conqueror* (1886).

4. This expression underpins research by ethologists and psychologists who consider that it might be as a response to the complexity of primates' social environment that their cognitive capacities became particularly developed, notably the art of dissimulating information from others for one's own profit. See R. Byrne and W. Whiten, *Machiavellian Intelligence: Social Expertise and the Evolution of Intellect in Monkeys, Apes, and Humans* (Oxford, UK: Clarendon Press, 1988).

5. Among living systems, crossover is produced during the constitution of each parents' reproductive cells and thus corresponds to a crossing of chromosomes belonging to the father or of those belonging to the mother, but not to crossings between chromosomes coming from both parents.

Chapter 11

1. David Williams, *Definitions in Biomaterials* (New York: Elsevier, 1987), 67.

Chapter 12

1. By Gerd Binnig and Heinrich Rohrer, who also were awarded the Nobel Prize.

Chapter 13

1. Manfred E. Clynes and Nathan S. Kline, "Cyborgs and Space," *Astronautics* (1960): 27–31.

2. Nancy Shute, *U.S. News & World Report*, July 23, 2006.

3. Ron, 73, in BBC news, 03/04/2009.

4. Professor Pierre Rabischong at Montpellier coordinates SUAW.

Conclusion

1. Janine M. Benyus, *Biomimicry: Innovations Inspired by Nature* (New York: Perennial, 2002).

2. Nevertheless, the American Society for the Prevention of Cruelty to Robots is ready to defend robots' rights!

3. See http://www.roboethics.org.

4. NewScientist Tech and AFP, 03/08/2007

5. In the story "Runaround" (*Astounding Science Fiction* [March 1942])—yet foreshadowed in earlier stories—Asimov wrote that (i) robots should not attack humans or allow humans to do them harm; (ii) robots should obey humans unless that conflicts with the first law; (iii) robots should ensure their own protection if that does not conflict with the other laws.

6. Joël de Rosnay, *2020, les scénarios du Futur* (Paris : Des Idées et des Hommes, 2007).

7. Candice Lombardi, CNET News, May 15, 2007.

8. *New Scientist* (2006), 192 (2578): 50.

Epilogue

1. Vercors, *You Shall Know Them*, translated by Rita Barisse (Boston: Little Brown, 1953), 224–225.

Further Information

Ayers, Joseph L., William J. Davis, and Alan Rudolph. *Neurotechnology for Biomimetic Robots*. Cambridge, MA: MIT Press, 2002.

Aylett, Ruth. *Robots: Bringing Intelligent Machines to Life*. Hauppauge, NY: Barron's Educational Series, 2002.

Bar-Cohen, Yoseph, ed. *Biomimetics: Biologically Inspired Technologies*. Boca Raton, FL: CRC Press, 2006.

Bar-Cohen, Yoseph, and Cynthia Breazeal. *Biologically Inspired Intelligent Robots*. Bellingham, WA: Spie Press, 2003.

Bekey, George A. *Autonomous Robots: From Biological Inspiration to Implementation and Control*. Cambridge, MA: MIT Press, 2005.

Benyus, Janine. *Biomimicry—Innovations Inspired by Nature*. New York: Perennial, 2002.

Breazeal, Cynthia. *Designing Sociable Robots*. Cambridge, MA: MIT Press, 2004.

Brooks, Rodney. *Cambrian Intelligence: The Early History of the New AI*. Cambridge, MA: MIT Press, 1999.

Brooks, Rodney. *Flesh and Machines: How Robots Will Change Us*. London: Vintage, 2002.

Chapuis, Alfred, and Edouard Gélis. *Le monde des automates*. Paris: Haraucourt, 1928. 2d ed. Geneva: Editions Slatkine, 1984.

Coineau, Yves, and Biruta Kresling. *Les inventions de la nature et la bionique*. Paris: Hachette Littérature, 1992.

Cordeschi, Roberto. *The Discovery of the Artificial: Behavior, Mind, and Machines before and beyond Cybernetics*. Boston: Kluwer, 2002.

Drexler, Eric. *Engines of Creation: The Coming Era of Nanotechnology*. New York: Anchor Books, 1986.

Floreano, Dario, and Claudio Mattiussi. *Bio-inspired Artificial Intelligence: Theories, Methods, and Technologies*. Cambridge, MA: MIT Press, 2008.

Forbes, Peter. *The Gecko's Foot: Bio-inspiration—Engineering New Materials and Devices from Nature*. New York: W. W. Norton, 2006.

Geary, James. *The Body Electric: An Anatomy of the New Bionic Senses*. New Brunswick, NJ: Rutgers University Press, 2002.

Gérardin, Lucien. *Bionics*. New York: McGraw-Hill, 1968.

Johnson, Franck E. *The Bionic Human: Health Promotion for People with Implanted Prosthetic Devices*. New York: Humana Press, 2006.

Kaplan, Frédéric. *Les machines apprivoisées: Comprendre les robots de loisir*. Paris: Vuibert, 2005.

Menzel, Peter, and Faith D'Aluisio. *Robo Sapiens: Evolution of a New Species*. Cambridge, MA: MIT Press, 2000.

Meyer, Jean-Arcady, and Agnès Guillot. Biologically inspired robots. In *Handbook of Robotics*, ed. Bruno Siciliano and Oussama Khalib. Berlin: Springer, 2008, 1395–1422.

Nachtigall, Werner, and Kurt G. Blüchel. *Das große Buch der Bionik*. Munich: Deutsche Verlags-Anstalt, 2000.

Nehaniv, Christopher L., and Kerstin Dautenhahn, eds. *Imitation and Social Learning in Robots, Humans, and Animals: Behavioural, Social, and Communicative Dimensions*. Cambridge, MA: MIT Press, 2007.

Nolfi, Stefano, and Dario Floreano. *Evolutionary Robotics: The Biology, Intelligence, and Technology of Self-Organizing Machines*. Cambridge, MA: MIT Press, 2001.

Perkowitz, Sidney. *Digital People: From Bionic Humans to Androids*. Washington, DC: Joseph Henry Press, 2004.

Vogel, Steven. *Cats' Paws and Catapults: Mechanical Worlds of Nature and People*. New York: W. W. Norton, 1998.

Webb, Barbara, and Thomas R. Consi. *Biorobotics: Methods and Applications*. Menlo Park, CA, and Cambridge, MA: American Association for Artificial Intelligence and MIT Press, 2001.

Wood, Gaby. *Edison's Eve: A Magical History of the Quest for Mechanical Life*. New York: Anchor Books, 2002.

Index